Lecture Notes in Intelligent Transportation and Infrastructure

Series Editor

The series "Lecture Notes in Intelligent Transportation and Infrastructure" (LNITI) publishes new developments and advances in the various areas of intelligent transportation and infrastructure. The intent is to cover the theory, applications, and perspectives on the state-of-the-art and future developments relevant to topics such as intelligent transportation systems, smart mobility, urban logistics, smart grids, critical infrastructure, smart architecture, smart citizens, intelligent governance, smart architecture and construction design, as well as green and sustainable urban structures. The series contains monographs, conference proceedings, edited volumes, lecture notes and textbooks. Of particular value to both the contributors and the readership are the short publication timeframe and the world-wide distribution, which enable wide and rapid dissemination of high-quality research output.

More information about this series at https://link.springer.com/bookseries/15991

Elżbieta Macioszek · Grzegorz Sierpiński
Editors

Present Approach to Traffic Flow Theory and Research in Civil and Transportation Engineering

Springer

Editors
Elżbieta Macioszek
Department of Transport Systems,
Traffic Engineering and Logistics
Silesian University of Technology
Gliwice, Poland

Grzegorz Sierpiński
Department of Transport Systems,
Traffic Engineering and Logistics
Silesian University of Technology
Gliwice, Poland

ISSN 2523-3440 ISSN 2523-3459 (electronic)
Lecture Notes in Intelligent Transportation and Infrastructure
ISBN 978-3-030-93369-2 ISBN 978-3-030-93370-8 (eBook)
https://doi.org/10.1007/978-3-030-93370-8

This Springer imprint is published by the registered company Springer Nature Switzerland AG
The registered company address is: Gewerbestrasse 11, 6330 Cham, Switzerland

Preface

Shopping centers, business areas, schools, hospitals and other separate areas of the city such as housing estates usually generate high values of traffic flows on the transport network. The vast majority of which are private cars. Traffic flows on the transport network are characterized by specific features, and they are dependent on each other. The effects of the movement of traffic flows are traffic congestion, air pollution, noise, low level of road traffic safety and parking problems. These effects are usually very onerous for people commuting to the above-mentioned traffic generators. In order to mitigate congestion and other unfavorable phenomena occurring in the transport network, various tools and methods are used to model traffic flows in transport networks.

The presented volume entitled "*Present approach to traffic flow theory and research in civil and transportation engineering*" provides an excellent opportunity to learn about the latest trends and achievements in the field of modern traffic flow theory as well as with challenges and solutions in the field of transport systems and traffic engineering. The volume is divided into four parts:

- Part 1. Civil and traffic engineering as the source of decision in transport planning,
- Part 2. Modelling and optimization as a support in the management of transport processes,
- Part 3. The safety issues in transport—human factors, applicable procedures, modern technology,
- Part 4. Structure and traffic organization in transport systems.

The content of the volume consists of selected papers submitted and presented at the 17th Scientific and Technical Conference "Transport Systems. Theory and Practice", organized by the Department of Transport Systems, Traffic Engineering and Logistics at the Faculty of Transport and Aviation Engineering of the Silesian University of Technology (Katowice, Poland). Each article has been reviewed by two independent reviewers (the so-called double-blind review). Among the topics discussed were issues related to current problems in the field of the present approach to traffic flow theory as well as transport systems. Numerous practical examples illustrate the possibilities of improving the current situation with a view to human welfare, sustainable development of transport systems and environmental protection.

Taking the opportunity, we would like to thank the authors for the submitted papers, for their great contribution to illustrating the numerous challenges posed by traffic flows in transport systems in the modern world, as well as for sharing the results of their scientific and research works. We also thank the reviewers for their insightful comments and suggestions that helped keep the volume high quality.

We also wish the readers a successful reading.

September 2021

Elżbieta Macioszek
Grzegorz Sierpiński

Organization

17th Scientific and Technical Conference "Transport Systems. Theory and Practice" (TSTP2021) is organized by the Department of Transport Systems, Traffic Engineering and Logistics, Faculty of Transport and Aviation Engineering, Silesian University of Technology (Poland).

Organizing Committee

Organizing Chair

Grzegorz Sierpiński	Silesian University of Technology, Poland

Members

Grzegorz Karoń	Silesian University of Technology, Poland
Ireneusz Celiński	Silesian University of Technology, Poland
Barbara Borówka	Silesian University of Technology, Poland
Agata Kurek	Silesian University of Technology, Poland
Marcin Staniek	Silesian University of Technology, Poland
Marcin J. Kłos	Silesian University of Technology, Poland
Adrian Barchański	Silesian University of Technology, Poland
Piotr Soczówka	Silesian University of Technology, Poland

The Conference took place under the Honorary Patronage

Ministry of Infrastructure
Marshal of the Silesian Voivodeship
Committee on Transport of Polish Academy of Sciences

Scientific Committee

Jacek Żak	Poznań University of Technology
Jolanta Żak	Warsaw University of Technology, Poland
Renata Żochowska	Silesian University of Technology, Poland

Referees

Ireneusz Celiński
Jacek Chmielewski
Piotr Czech
Boris Davydov
Elżbieta Macioszek
Marcin Staniek
Grzegorz Sierpiński
Agnieszka Tubis
Grzegorz Wojnar

Contents

Civil and Traffic Engineering as the Sources of Decision in Transport Planning

Tram and Bus Stops in Cities – An Overview of Assessment Methods

Mateusz Rydlewski(✉)

Wroclaw University of Science and Technology, Wrocław, Poland
mateusz.rydlewski@pwr.edu.pl

Abstract. Tram and bus stops play a key role in the functioning of public transport systems. Their condition, functionality and assured safety level are particularly important from the point of view of the assessment of this infrastructure element. The article presents an overview of the methods of assessing stops, the classification of which is based on the defined 5 distinguishing factors of tram and bus stops infrastructure assessment. The high availability of assessment methods from various cities around the world indicates a significant problem related to the possibility of obtaining information on the level of fulfillment of given requirements in a given assessment area. The conducted review shows many similarities between different methods, but also points to areas where certain solutions are lacking in terms of, for example, customization of assessment methods.

Keywords: Public transport · Safety · Functionality · Accessibility · Tram stops · Bus stops · Assessment method

1 Introduction

The stop infrastructure is an integral part of any public transport network in cities. The basic element of each stop is a space separated in the road lane, where it is possible to wait for the arrival of the vehicle serving the stop. An integral part of each stop is the stop post with a sign informing about the presence of the stop and the timetable. Its main purpose is to convey information to different road users. For travelers, it is information about indicating the place where the vehicle will stop and access to the timetable of vehicles at the stop. For drivers, the stop sign determines the need to be particularly careful and for public transport vehicle drivers it indicates the place where the vehicle should be stopped and passengers replaced. The remaining basic elements of the stop's equipment also include [1]: a garbage can, a bench and a shelter for the bus stop. Depending on the needs, the stops can be equipped with additional elements of equipment, such as: lighting poles, monitoring cameras, dynamic passenger information infrastructure, green areas, facilities for people with disabilities and others. The presence of additional elements of equipment may affect many factors, such as the comfort and functionality of using the stop, accessibility for people with disabilities or, as shown in the research [2], reducing the nuisance of waiting for a vehicle.

The review of assessment methods applies to tram and bus stops, which can be divided into six basic types:

E. Macioszek and G. Sierpiński (Eds.): TSTP 2021, LNITI, pp. 3–20, 2022.
https://doi.org/10.1007/978-3-030-93370-8_1

- island stops – Fig. 1 – a type of stop where its space is separated in a road lane and is located between the lane of a public transport vehicle and the lane or lanes for other vehicles. This stop is most often intended for the service of trams whose track is located in the axis of the road, but can also be served by buses that run in a separate lane in the axis of the road,
- stops at the sidewalk – Fig. 2 – a type of stop where the waiting area covers a section of the pavement and the vehicle stops directly at the curb to serve passengers. Most often used by buses, but it is also used in the case of tram stops.
- stops in the bay – Fig. 3 – the type of bus stop where the vehicle stops at the edge of the stop, often on the footpath. The bus is stopped in the bay, so it is possible for other vehicles to bypass the bus during the exchange of passengers,
- stops with access to vehicle in the lane area – Fig. 4 – a type of tram stop where the exchange of passengers takes place in the area of the lane for vehicles and the entrance to the vehicle takes place directly from the road. In the vicinity of such a stop, additional measures to improve safety are often applied, such as traffic lights preventing the entry of vehicles during passenger exchange,
- Vienna type tram stops – Fig. 5 – a type of stop (mainly a tram stop) in which the exchange of passengers takes place in the area of the lane, which is raised to the level of the sidewalk, so that the exchange of passengers is carried out similarly to the case described above, with the difference that the entrance to the vehicle takes place from the elevation of the road. Lane elevation is also a form of vehicle speed reduction resulting from the need to climb,
- stops in the form of an anti-bay – Fig. 6 – a type of stop intended mainly for trams but also buses. In the vicinity of this type of stop, public transport vehicles most often move along a separate lane and in the area of the stop itself there is a narrowing of the road, the connection of the lane for passenger vehicles and public transport, and the extension of the walkway to the edge of the stop.

Stops, even of the same type, differ from each other, often quite significantly and it is influenced by many factors. The main elements determining the differences are: the type of supported vehicle or vehicles, the traffic volume of public transport and passenger vehicles, the location of the stop in the city area, the size of the passenger exchange or infrastructure limitations. Stops are a very important element of the efficient and safe functioning of public transport in cities. Therefore, it is very important to ensure an appropriate level of safety and functionality of the stops. However, the multitude of solutions encountered at stops determines the need to look at the problem of evaluating stops in a broad sense. In the further part of the chapter, the identified methods of evaluating stops are described, the subject and method of evaluation are presented, and on the basis of the conducted analysis, a list of the most frequently appearing evaluation methods is prepared. The identification of the assessment methods was based on Polish and foreign research from various places in the world. The analyzed methods mainly related to the assessment of tram stops, some of them also included the assessment of the bus stop infrastructure.

Fig. 1. Island tram stop

Fig. 2. Tram stop on the sidewalk

Fig. 3. Bus stop in the bay

Fig. 4. Tram stop with access from the road

Among Polish studies, the methods for which case studies were presented in the following cities and agglomerations were analyzed: Gdańsk, Upper Silesian Conurbation, Kraków, Olsztyn, Poznań, Szczecin, Warsaw, Wrocław. This is an overview of the methods for assessing stops in most large Polish cities and more than half of those

Fig. 5. Viennese tram stop.

Fig. 6. A tram stop with an anti-bay

with a tram network. On the other hand, the most important urban centers where the methods of evaluating stops were analyzed include: Australia (Melbourne, Monash, Adelaide or Queensland), Italy (including Rome, Bologna), Spain (including Barcelona, Valencia, Seville), USA (Portland, Las Vegas, Los Angeles and Florida) or Sweden (including Malmo, Goteborg, Lund) and other cities in Europe and Asia.

2 Classification of Evaluation Methods

This chapter presents the criteria for the division of the stop infrastructure assessment, identified on the basis of the review of the methods. This division was defined on the basis of the determination of five main factors determining the differences in the analyzed assessment methods. The division is presented in Fig. 7 and their detailed description is presented in the following subsections.

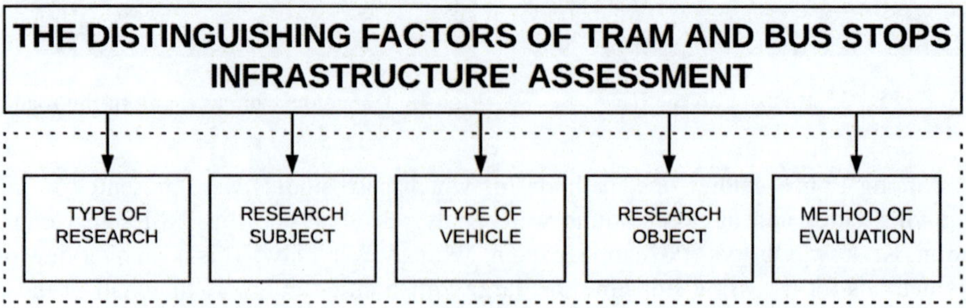

Fig. 7. The distinguishing factors of tram and bus stops infrastructure' assessment

2.1 Type of Research

Various evaluation methods can be identified in the literature, but they are always strictly dependent on the nature of the identified research problem [3]. The basic and most important division results directly from the evaluation method, in which we distinguish quantitative and qualitative evaluations [4]. The former often has strictly defined evaluation parameters, and the evaluation process is independent of the person making the measurement. It is a quantitative research in which the researcher determines the values of the variables [3]. The second approach, those of a more subjective nature, refers to the qualitative assessment, which, in the absence of properly defined assessment parameters, may have different results depending on the assessor.

Among the methods are those in which there is a combination of qualitative and quantitative assessment methods. Combining quantitative and qualitative methods in a common and integral way of assessment is common in many scientific areas. Combining assessment methods can be found in a number of studies of a social or psychological nature [3, 4] but also in the technical industry such as construction [5] or transport [6] or in the area of decision-making processes [7]. The factors determining the choice of the evaluation method are the aim and research problems as well as the characteristics of the research object [3].

Based on the conducted review of methods and type of test, three main types of test were identified:

- quantitative research,
- qualitative research,
- mixed (quantitative and qualitative research).

An example of quantitative research may be the analysis of the number of accidents that took place in a given area or with a given type of facility. The specificity of research related to the stop infrastructure, where the number of accidents is lower than in the case of e.g. road accidents, determines the use of mixed types of research, which is common in this type of research [8].

Detailed information on the classification of selected methods for the identified types of research is presented later in this chapter.

2.2 Research Subject

The basic and the most frequent subject of research in the field of the operation of the tram and bus network is the aspect of broadly understood safety, which, according to [9], has the highest priority in terms of increasing the level of public transport services. The aspect of public transport safety can be found in many studies in Poland (e.g. [10–12]) and other countries such as Switzerland [8], France [13], Australia [14, 15] Sweden [16], China [9], USA [17], Italy [18, 19]. The safety aspects discussed in the research relate to various safety-related spheres.

The first identified group of studies are those directly related to the *safety of public transport users*. These studies take into account the impact of various factors on the safety of public transport users. In Poznań [12], the focus was on identifying the sources of threats at tram stops. Another Polish study in Gdańsk [10] covered several types of research on the infrastructure of the tram network, which had a significant impact on the safety of its users. On the other hand, in Melbourne [14], a comparative analysis of safety in the vicinity of stops was performed on solutions improving the accessibility of the stop by its users. Research carried out in Gotehenburg [20] shows that most accidents involving a tram take place in the area or in the immediate vicinity of a tram stop.

The second group of identified types of safety research were those related to the *safety of other road users*. Based on the analysis of accidents in Melbourne, research [15] was carried out, on the basis of which dangerous places in the tram network were identified. In Wuhan (China) [9] a study was conducted related to the perception of bus drivers and their exposure to an accident, the elements of the study were a survey among drivers and an analysis of accident statistics. French research [13] also included an analysis of accidents involving public transport vehicles, on the basis of which dangerous places were diagnosed. In Gdańsk [10], the entire tram network was analyzed and one of the areas of the study was the identification of conflict places for different users. In turn, Australian research [21] focused on the places of potential risk between motor vehicles and trams.

The last group of studies identified in the area of safety were those relating to *the sense of personal safety*. [16] focused on research related to the sense of safety by users of the Stockholm metro. On the other hand, the research in Melbourne (Australia) [17] concerned an application aimed at obtaining information about the perception of personal safety by users in different places and times. Research in the USA [18] was carried out through surveys, the respondents of which were users of public transport stops.

In [14], Currie states that tram stops pose a challenge in terms of safety, accessibility and transport efficiency. The context of transport efficiency in the aspect of research on stop infrastructure was considered irrelevant. On the other hand, the research on *accessibility to tram stops* is widely used, especially in the area of GIS analyzes [22, 23].

The last group of the subject of the research are those related to the evaluation of the *functionality of the stop infrastructure*. The research conducted in Poznań [24] was aimed at assessing the functionality of tram stops. The studies [25] and [26] focused on the equivalent assessment of the safety and functionality of tram loops based on the assessment criteria indicated in the method.

An important element related to the functionality and accessibility of stops is the aspect related to *people with disabilities*. It was discussed in research carried out in England [27], where it was shown that transfer nodes have a greater problem with accessibility than the public transport fleet.

Figure 8 shows a detailed division of the *research subject* that can be identified in the research on the tram and bus stop infrastructure.

2.3 Type of Vehicle

The article focuses on the review of the methods of assessing the stop infrastructure in the urban area, therefore the division by vehicle type is strictly determined by the types of vehicles used in public transport in the urban area. With such limitations, the following modes of transport have been identified:

- tram,
- bus,
- metro,
- train,
- water tram.

However, due to the significant differences between the stops of the above-mentioned means of public transport, the review of methods for assessing the stop infrastructure has been limited to the first two items. However, due to the specificity of the traffic of both trams and buses, the division in the area of *the type of vehicle* servicing the stop has been extended to a common stop, i.e. a tram-bus stop, where there is a stop and replacement of passengers from both types of vehicle.

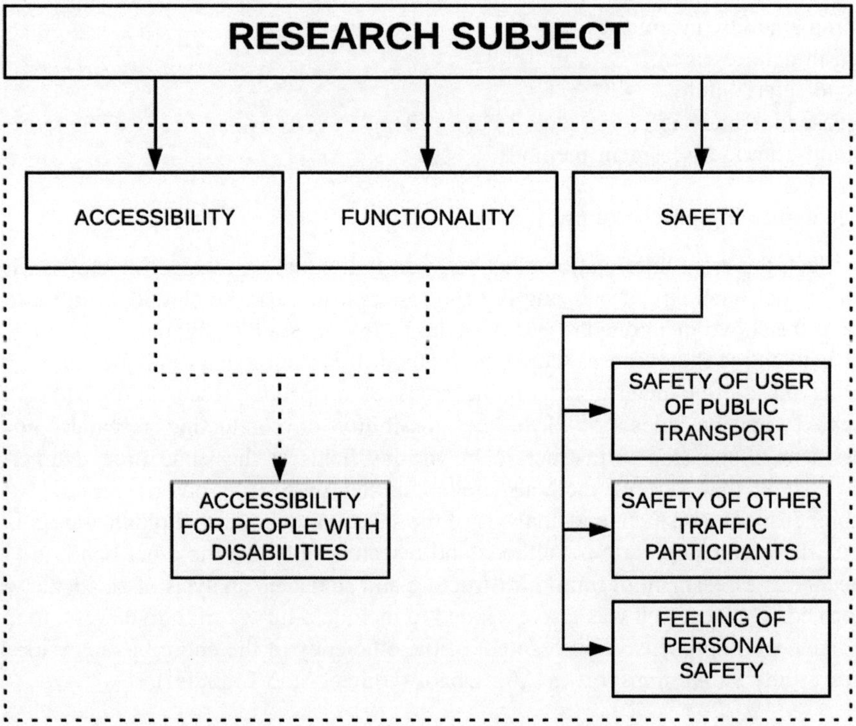

Fig. 8. Detailed breakdown of *research subject* factor, source: own works

2.4 Research Object

As part of this criterion determining the method of assessing the stop infrastructure, two types of facilities have been distinguished that can be tested, and these are:

- real object,
- designed object.

In this case, similarly to the type of vehicle, an additional group was taken into account – *real or projected object*. Selected universal methods allow for the assessment of the object regardless of its condition. The applied division allows for the classification of assessment methods based on any eventuality resulting from the presence of a stop, and also allows for the assessment of planned stop solutions. Few research methods make it possible to evaluate both the real and the designed object, in [24] a method combining such possibilities was presented.

2.5 Method of Evaluation

The last of the criteria for dividing the assessment of the tram and bus stop infrastructure concerns *method of evaluation*. Based on the literature review, 7 main groups differentiating the evaluation method were distinguished, and they are as follows:

- simple audits, inventories,
- crash-data research,
- road safety audit,
- geospatial analysis,
- multi-criteria assessment methods,
- user rating,
- accompanying observations.

In fact, there are interrelationships between individual assessment methods, which result in such methods of carrying out the research that can be classified into two or more of the above-mentioned research methods. In each such study, the most important thing is to define the purpose, scope and method of measurement implementation. The latter is most often determined on the basis of a specific research objective. The presented division does not close the possibility of conducting research on the assessment of the stop infrastructure in various fields at the same time. Numerous studies show that various methods are combined with each other, e.g. in [10], a comprehensive assessment and analysis of the safety of existing and planned tram lines was carried out, using various methods and research tools. On the other hand, in [11], the technical assessment of tram infrastructure and statistical analysis of accidents were performed. This research was also extended to include a survey of tram drivers. In turn, [28] made a comprehensive assessment of the efficiency of the entire bus network that makes up the public transport in Abu Dhabi (United Arab Emirates).

2.5.1 Simple Audits, Inventories

The first method of assessing the stop infrastructure is simple audits carried out by an observer. These tests are most often characterized by the use of formalized measurement sheets, on which the auditor marks the appropriate answers depending on the actual state. This survey may include several major rating indicators to which weights can be assigned. This type of research is usually not very time-consuming.

In the study conducted in Olsztyn [29], evaluation indicators were used, which were obtained on the basis of surveys conducted on a group of 100 specialists from the public transport industry. Among the evaluation indicators, 24 elements were identified and identified during the field visit. The final results classify the stop into one of the four stop classes proposed by the authors.

Another example of a survey that is a form of inventory is [30–33]. In this study, a very detailed inventory card was proposed for the assessment of the surface condition of bus stops. One of the test elements was also the assessment of the surface in the bay area.

2.5.2 Crash-Data Research

A very popular method of safety assessment is the analysis of historical data based on accidents involving a specific group of users or in a specific area of the site. This type of research is of little use in the case of the stop infrastructure due to the relatively small number of accidents occurring in the stop area. This type of research is definitely more widely used in the case of a more comprehensive approach to the assessment of public transport safety, e.g. in a metropolitan area [34]. In this case, due to the large area of the study, it is possible to obtain more data. For example, in the study [31] from the period of three years, data on over 4 500 collision and non-collision incidents in the Portland Oregon metropolitan region. A similar approach was used in [35] where the level of pedestrian safety was determined in the vicinity of bus stops in Florida. Similarly, the level of safety was determined on the basis of the crash data in Kolkata [36].

Often, for a comprehensive approach to assessment, apart from the accident data itself, a combination of this method with spatial analyzes is used, such as in [23] or for analyzes using an advanced before-after crash analysis approach [32]. This study analyzed the change of the type of tram stop, the elevation of which in the area of the waiting area resulted in an increase in the safety level.

2.5.3 Road Safety Audit

This form of conducting the bus stop infrastructure survey is the most formalized method of carrying out the survey in Poland. A road traffic safety audit is an example of an activity strictly regulated by law. Persons authorized to implement it are properly trained and experienced, which is confirmed by a passed state examination. A road safety audit is often outsourced when numerous or catastrophic accidents are identified. The scope of such an audit is comprehensive and the assessment of the stop itself is usually only a fragment of a comprehensive look at a given area of an intersection, road or otherwise limited area.

2.5.4 Geospatial Analysis

GIS (geographic information system) tools and appropriate software necessary to perform relevant analyzes are most often used to carry out this type of research. The analyzes themselves are carried out on the basis of input data that must be defined and specified in the research process. Spatial analyzes in the context of public transport stops are most often used to determine the pedestrian accessibility to a given stop. Isochrones are used for this. Lines that mark an area with the same, set, time reachability from a selected point, in this case a stop. The band method and circular buffer method can also be used to determine availability. The description of both methods with a comparative analysis is presented in [33]. These studies show that the streak method is more accurate.

Another application of GIS tools in the aspect of spatial analyzes is to determine the level of exposure to previously identified hazardous events, ie collisions and accidents, eg. [23]. This method of carrying out the research is an example of combining this method with the one described in point 2.5.2. Spatial analyzes may also include safety aspects determined on the basis of the location of the stop and selected adjacent infrastructure elements of specific nature [35]. Another example of combining research methods with each other is [37] where spatial analyzes were combined with a widely conducted survey among public transport users.

2.5.5 Multi-criteria Assessment Method

This research method is very often a much more developed audit carried out in a manner similar to that described in point 2.5.1. The main difference between these methods results from the classification of selected evaluation criteria into a specific index, which can then be a component to determine the class of objects or, for example, the level of service, depending on the selected evaluation parameters. Grouping the assessment criteria into specific indicators is most often thematically so that the indicator shows the assessment in a specific assessment area defined by the author of the study. In the research [38], 8 indicators can be distinguished within which transfer nodes are assessed, including 5 indicators relating directly to the functionality and safety of stops. These studies are a continuation of research conducted by researchers from Warsaw [39]. The authors of these studies also developed a multi-criteria method of assessing transfer nodes for the capital of Poland.

In this type of research, two research subtypes can be distinguished that determine the *use of weights* (e.g. [40, 41] or *no use of weights* [25, 42, 43]. they are most often selected on the basis of surveys or the Delphi method. Depending on the authors, different types of respondents selected to determine the weights of individual criteria were observed. Most often, however, in the field of research on the safety of public transport stops, people who have a real impact on determining the weights are people with experience in managing, designing or maintaining public transport. The knowledge of experts to determine the weights is widely used in the field of research on the discussed issues [41, 44]. Sometimes the opinions of the stop users themselves are also used to determine the weights, although this form is more applicable in research related to functionality and assessment of the perception of selected aspects by users than in terms of safety. However, the determination of weights by experts also takes place in relation to research on functionality and accessibility for people with disabilities [40]. In [45], the research concerned the adaptation of infrastructure to people with reduced mobility, and the weights of the study criteria in this area were selected based on surveys conducted among people with disabilities.

Multi-criteria studies often allow to assess the level of service, security or functionality, depending on the purpose of the study. Quantitative research was conducted in [46] to determine the level-of-safety. On the other hand, [19] used an IT model to assess the level of safety.

2.5.6 User Rating

The research method presenting subjective assessments of users of public transport stops is a simple study in which a questionnaire or an interview with users is most often used to determine the assessment [47]. Research of this type indicates the perception of the reception of specific solutions and may be particularly useful in assessing the effectiveness of using a given solution in the area of a stop. This research is an assessment or indication of the user's expectations. They can be implemented to assess two basic aspects: *the perception of personal safety of stop users* [48] or *the assessment of the stop infrastructure* [49]. The assessment related to the infrastructure relates to aspects of functionality, often including the convenience of using a given stop. The aspects of security assessment are implemented here solely in the area of users' perceptions. The method makes it impossible to determine the safety of the stop itself.

2.5.7 Accompanying Observations

This survey can be carried out in several ways. The observation can be made directly by an observer standing near the research facility or through, for example, video recording with the use of cameras placed on masts, road signs, poles or elements of traffic light equipment in the vicinity of the stop. The use of drones recording the stop area is also becoming more and more common in this type of research. Observation with the use of cameras (stationary and those on a drone) can be made in real time or played and processed by an observer or computer software. The accompanying observation can be used in two areas, which are: *identification of places of collision between different road users* and *analysis of the behavior of road users* [50]. In the first case, it is crucial to identify specific points that are places of collision. This type of survey is especially used for tests at interchange nodes with several stops. The analysis of the behavior of traffic participants can also be used in the case of research on transfer nodes in order to, for example, determine the traffic pattern in the area of a node, although this form of accompanying observation is also applicable in the area of stops. In [51], the impact of traffic organization on the behavior of pedestrians in the area of pedestrian crossings with tram tracks was analyzed. In this type of research, it is possible to determine, for example, the use of elements of the stop infrastructure, the way passengers move when passengers are replaced, which allows to identify dangerous places and possible bottlenecks in the area of the stop under analysis.

3 Detailed Classification of Scoring Infrastructure Assessment Methods

The identified methods of assessing the stop infrastructure are characterized by different approaches to the implementation of the survey.

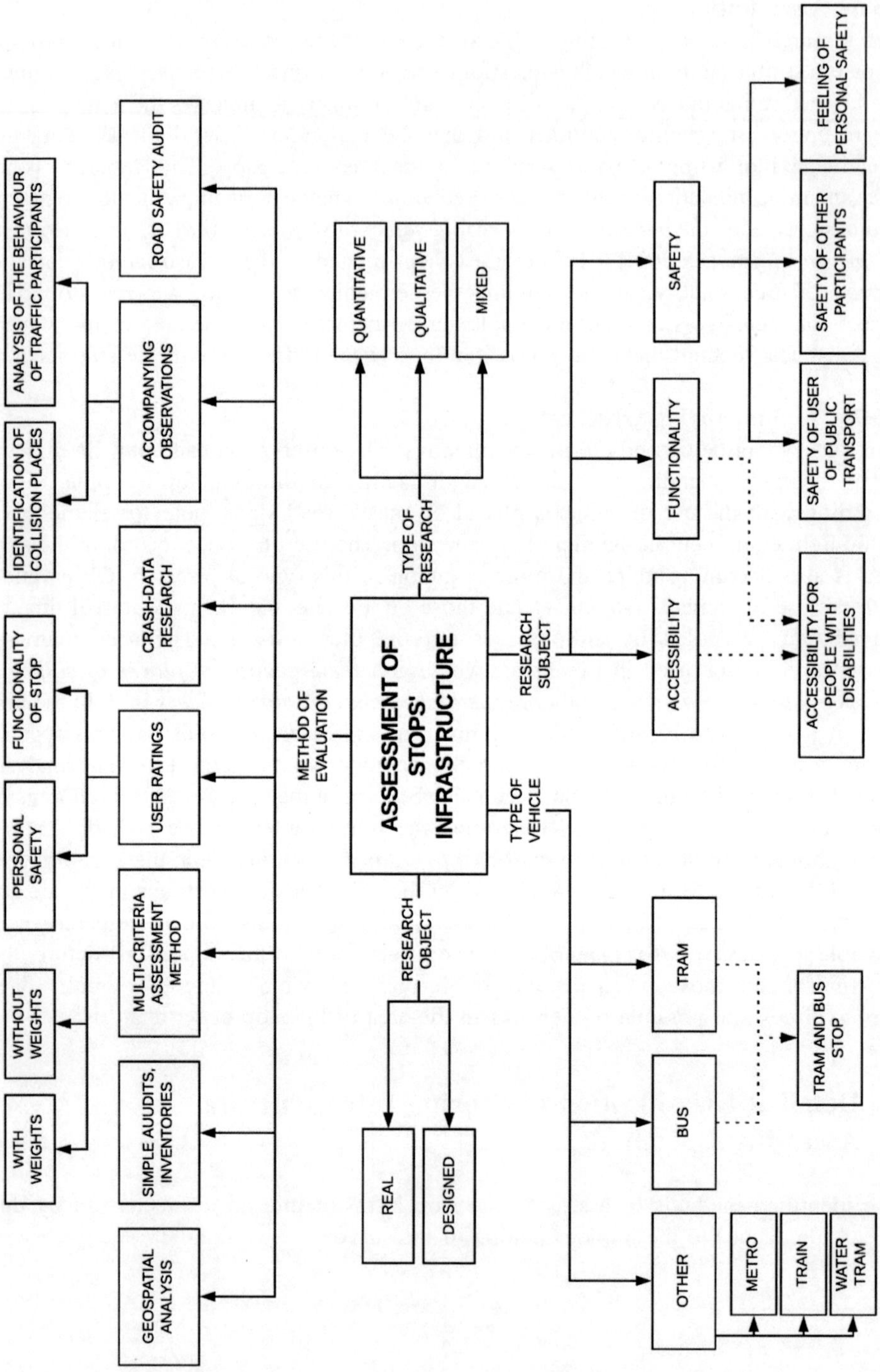

Fig. 9. Detailed breakdown of the assessment methods of stops' infrastructure

Among the presented methods, there are often those that are characterized by a multi-area research method, under which research is carried out using various forms. Figure 9 shows the division of methods for assessing the stop infrastructure based on 5 classifying factors identified at the beginning of this chapter.

Table 1 presents selected methods of assessing the stop infrastructure from all over the world. These data include the basic categorization of individual methods, the designations used in the table are described in Fig. 10.

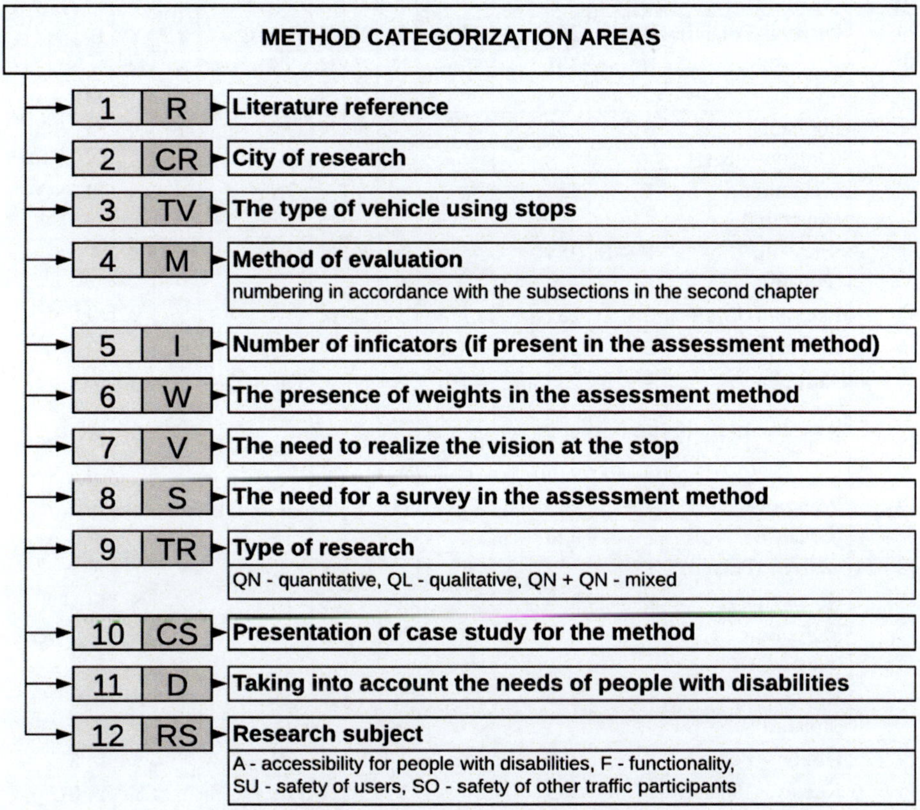

Fig. 10. Method categorization areas

Table 1. Detailed information on the identified evaluation methods.

R	CR	TV	M	I	W	V	S	TR	CS	D	RS
1	2	3	4	5	6	7	8	9	10	11	12
1	Poznań (Polska)	T + B	251	8 (29)	Y	Y	N	QN	Y	Y	SU
12	Poznań (Poland)	T	251	6	N	Y	N	QL + QN	Y	N	SU
19	Bolonia (Italy)	B	255	120	Y	Y	N	QL + QN	Y	N	SU

(*continued*)

Table 1. (*continued*)

R	CR	TV	M	I	W	V	S	TR	CS	D	RS
1	2	3	4	5	6	7	8	9	10	11	12
23	Adelajda (Australia)	B	252 and 254	–	N	N	N	QN	Y	N	SU
24	Poznań (Poland)	T	251	6(43)	N	Y	N	QN	Y	Y	F, SU
25 and 26	Wrocław (Poland)	T	255	8	N	Y	N	QL + QN	Y	Y	F, SU
29	Olsztyn (Poland)	B	251	24	Y	Y	N	Q + Q	Y	Y	F
30	–	B	251	50	N	Y	N	QN	N	N	F, SU, SO
31	Portland (Oregon–USA)	B	252	–	–	N	N	QN	Y	N	SU SO
32	Melbourne (Australia)	T	252	–	–	Y	N	QN	Y	N	SU SO
33	Szczecin (Poland)	T + B	254	–	N	N	N	QN	Y	N	F
35	Floryda (USA)	B	252	17	Y	N	N	QN	Y	Y	SU, SO
36	Kalkuta (Indie)	B	252	4 (30)	Y	Y	N	QL + QN	Y	N	SU
37	Australia	T + B	254 and 256	4 (9)	N	N	Y	QL + QN	Y	N	F
38	Kraków (Poland)	T + B	255	8 (13)	N	Y	Y	QL + QN	Y	Y	F, A, SU, SO
39	Warszawa (Poland)	T + B	255 and 256	8 (42)	N	Y	Y	QL + QN	Y	Y	F, SU, SO, A
40	Olsztyn (Poland)	T	2551	10	Y	Y	N	QL	Y	Y	F, A
41	Rome (Italy)	B	255	7	Y	Y	N	QN	Y	N	F
42	Melbourne (Australia)	T	255	4 (39)	N	Y	N	QL + QN	Y	N	SU, SO
43	Wrocław (Poland)	T	255	3	N	Y	N	QN	Y	N	SU, SO
44	Olsztyn (Poland)	B	255	15	Y	Y	N	QN	Y	Y	F, SU, A
45	Kraków (Poland)	T + B	255	8	Y	Y	N	QL + QN	Y	Y	F, SU, A
46	China	B	255	7 (17)	Y	Y	N	QN	Y	N	SU
47	Los Angeles (USA)	B	256	5 (46)	N	N	Y	QL	Y	N	F
48	Melbourne (Australia)	T + B	256	10	N	N	Y	QL	Y	N	SU
49	Santander (Spain)	B	256	6	N	N	Y	QL	Y	N	F
50	GZM Metropolis (Poland)	T	257	–	–	Y	N	QN	Y	N	SU, SO

4 Summary

Stops are a very important element of the public transport network. The variety of stop types, their location and often infrastructure constraints pose the challenge of creating safe and functional places for passenger exchange. The assessment methods collected and discussed in the article, despite many, often very large, differences between them, have a common goal. It is the determination of the safety measure or functionality of the stop. These assessments make it possible to determine e.g. the prioritization of taking improvement measures aimed at ensuring safe and comfortable use of public transport. The wide access to evaluation methods allows us to state that there is a wide possibility of evaluating the stops.

A significant part of the identified methods are those in which the assessment is carried out through a multi-criteria approach. Determining the significance levels of certain assessment indicators in all methods is done using the knowledge and experience of people involved in public transport, including stop infrastructure. Surveys conducted among city experts, or the users themselves, make it possible to determine the weightings of the significance of selected elements of the assessment. In most of the identified research methods, the weights of individual criteria were determined with the participation of experts in a given field.

Research on the assessment of the stop infrastructure is conducted all over the world, which allows us to conclude that this is an area of development in terms of research. On the basis of the identified methods, it can be concluded that most of them are qualitative or mixed. Many evaluation methods, even those that do not classify to multi-criteria evaluation methods, have numerous indicators and criteria in their method. This solution allows for grouping specific elements into one scope of assessment. There is also a noticeable trend of including the needs of people with disabilities in newer assessment methods. When analyzing the publication dates of individual methods, it can be observed that the youngest methods contain the most elements related to the needs of people with disabilities. As in the case of the needs of the disabled, over the years, an increasing level of research advancement and the number of areas studied within one research method can be noticed.

In the discussed methods, there is a certain gap in terms of the customization of the evaluation method. The methods make it possible to evaluate various types of stops without preconditioning the scope of the test and the lack of selection of parameters strictly assigned to a specific type of facility. This creates the need to explore the subject of customization of assessment methods depending on the type of object in order to present a method that, being a universal assessment tool, would be fully adapted to the nature of the assessed object.

References

1. Koperski, Ł., Zmuda-Trzebiatowski, P.: Ocena stanu infrastruktury przystankowej. Studium przypadku: Poznański Łazarz. In: Prace Naukowe Politechniki Warszawskiej, Transport, vol. 12, pp. 181–190 (2018)
2. Nobuaki, O., Takashi, O., Shuichi, M., Kenji, O., Noboru, H.: Passengers' waiting behavior at bus and tram stops. In: 6th International Conference of Traffic and Transportation Studies Congress (ICTTS). Nanning (China) (2008). https://doi.org/10.1061/40995(322)48
3. Jasiński, M., Kowalski, M.: Fałszywa sprzeczność: metodologia jakościowa czy ilościowa? In: Haber, A. (ed.) Ewaluacja ex-post. Teoria i praktyka badawcza, pp. 94–114. Polska Agencja Rozwoju Przedsiębiorczości, Waszawa (2007)
4. Urbaniak-Zając, D.: On combining qualitative and quantitative research – expectations and doubts (in Polish). Educ. Stud. Rev. **26**, 121–138 (2018)
5. Hoła, B.: Qualitative and quantitative modelling of accident situation in building industry (on Polish). In: Prace Naukowe Instytutu Budownictwa Politechniki Wrocławskiej. Oficyna Wydawnicza Politechniki Wrocławskiej, Wrocław (2008)
6. Stokłosa, J.: Risk management in transport chains (in Polish). In: Zeszyty Naukowe WSEI. Seria Transport i Informatyka, vol. 1, no. 1, pp. 89–98 (2011)
7. Książek, M.: Use of the multicriteria methods for assessment of investment in decision-making process (in Polish). Logistyka **3**, 1429–1438 (2011)
8. Marti, C.M., Kupferschmid, J., Schwertner, M., Nash, A., Weidmann, U.: Tram safety in mixed traffic: best practices from Switzerland. J. Transp. Res. Board **2540**(1), 125–137 (2016)
9. Ma, M., Yan, X., Huang, H., Abdel-Aty, M.: Safety public transportation occupational drivers: risk perception, attitudes, and driving behavior. J. Transp. Res. Board **2145**(1), 72–79 (2010)
10. Budzyński, M., Szmagliński, K., Jamroz, K., Birr, K., Grulkowski, S., Wachnicka, J.: Assessing tram infrastructure safety using example of the city of Gdańsk. J. KONBiN **49**(3), 293–322 (2019)
11. Szmagliński, J., Grulkowski, S., Birr, K.: Identification of safety hazards and their sources in tram transport. MATEC Web Conf. **231**, 1–9 (2018)
12. Lingo, K., Smoczyński, P., Kominowski, J., Motyl, M.: Identyfication of the sources of hazards at tram stops (in Polish). Transp. Miejski i Regionalny **9**, 31–35 (2020)
13. Blancheton, M., Fontaine, L.: Tram safety management – french experience. In: Transportation Research Board 93rd Annual Meeting, pp. 1–16. Washington (2014)
14. Currie, G., Smith, P.: Innovative design for safe and accessible light rail of tram stops suitable for streetcar-style conditions. J. Transp. Res. Board **1995**(1), 37–46 (2006)
15. Andreassend, D.C., Donnelly, P.G.: Accidents to tram patrons. In: 18th Australian Road Research Board Conference, vol. 11, pp. 172–181 (1982)
16. Ceccato, V.: Moving safely. Crime and Perceived Safety in Stocholm's Subway Stations, Lexington Books, Maryland (2013)
17. Hamilton, M., Salim, F., Cheng, E., Choy, S.L.: Transafe: a crowdsourced mobile platform for crime and safety perception management. ACM SIGCAS Comput. Soc. **41**(2), 32–37 (2011)
18. Morse, L.B., Benjamin, J.M.: Analysis of feeling of security on public transit among residents of small urban area. J. Transp. Res. Board **1557**(1), 28–31 (1996)
19. Amadori, M., Bonino, T.: A methodology to define the level of safety of public transport bus stops, based on the concept of Risk. Procedia Soc. Behav. Sci. **48**, 653–662 (2012)

20. Hedelin, A., Bjornstig, U., Brismar, B.: Trams – a risk factor for pedestrians. Accid. Anal. Prev. **28**(6), 733–738 (1996)
21. Candappa, N., Corben, B., Yuen, J.: Addressing the conflict potential between motor vehicles and trams at cut-through locations. In: Technical Report No. 317. Monash University Accident Research Centre, Melbourne (2013)
22. Yuksekol, I.: Evaluation of Pedestrian Safety Around Bus Stops Using Geographic Information Systems – Dissertation. M.S. – Master of Science. Middle East Technical University, Ankara (2012)
23. Troung, L.T., Somenahalli, S.V.C.: Using GIS to identify pedestrian-vehicle crash hot spots and unsafe bus stops. J. Public Transp. **14**(1), 99–114 (2011)
24. Szaumkessel, D., Winiewicz, M., Dahlke, G.: Criteria for the evaluation and design of tram stops (in Polish). Logistyka **4**, 1282–1294 (2014)
25. Tubis, A., Rydlewski, M., Budzyński, M.: The indicator assessment of safety and functionality of tram loops. J. KONBiN **50**(3), 21–41 (2020)
26. Tubis, A., Rydlewski, M., Budzyński, M.: Safety and functionality assessment of tram loops in Wroclaw. J. KONBiN **51**(1), 159–172 (2021)
27. Ferrari, L., Berlingerio, M., Calabrese, F., Reades, J.: Improving the accessibility of urban transportation networks for people with disabilities. Transp. Res. Part C **45**, 27–40 (2014)
28. Hassan, M.N., Hawas, Y.E., Ahmed, K.: A multi-dimensional framework for evaluating the transit service performance. Transp. Res. Part A **50**, 47–61 (2013)
29. Podciborski, T., Mikluszka, P.: Assessment of the development status of municipal public transport bus stops on the example of the city of Olsztyn (in Polish). Logistyka **4**, 5335–5339 (2015)
30. Sołowczuk, A., Gardas, P.: The proposed assessment method of pavement condition at bus bays and bus stops. Roads Bridges **12**, 425–439 (2013)
31. Strathman, J.G., Wachana, P., Callas, S.: Analysis of bus collision and non-collision incidents using transit ITS and other archived operations data. J. Safety Res. **41**, 137–144 (2010)
32. Naznin, F., Currie, G., Logan, D., Sarvi, M.: Safety impacts of platform tram stops on pedestrians in mixed traffic operation: a comparison group before-after crash study. Accid. Anal. Prev. **86**, 1–8 (2016)
33. Kaszczyszyn, P., Sypion-Dutkowska, N.: Walking access to public transportation stops for city residents. A comparison of methods. In: Sustainability, vol. 11, no. 3758 (2019)
34. Ulak, M.B., Kocatepe, A., Yazici, A., Ozguven, E.E., Kumar, A.: A stop safety index to address pedestrian safety around bus stops. Saf. Sci. **133**(105017) (2021)
35. Hazaymeh, K.: GIS-based safety bus stops – Serdang and Seri Kembangan case study. J. Public Transp. **12**(2), 39–51 (2009)
36. Badland, H.M., Rachele, J.N., Roberts, R., Giles-Cort, B.: Creating and applying public transport indicators to test pathways of behaviorus and health through an urban transport framework. J. Transp. Health **4**, 208–215 (2017)
37. Bryniarska, Z., Zakowska, L.: Multi-criteria evaluation of public transport interchanges. Transp. Res. Procedia **24**, 25–32 (2017)
38. Olszewski, P., Kurkowski, P., Krukowska, H., Penar, K.: Use of the method indicator and survey for the evaluation of selected interchange node in Warsaw. Biuro Drogownictwa i Komunikacji Urzędu m.st. Warszawy, Warsaw (2011)
39. Podciborski, T.: A method for evaluating tram stops based on passenger expectations and the needs of disabled persons. In: 10th International Conference, Environmental Engineering. Lithuania (2017). https://doi.org/10.3846/enviro.2017.115
40. Corazza, M.V., Favaretto, N.: A methodology to evaluate accessibility to bus stops as a contribution to improve sustainability in urban mobility. Sustainability **11**(3), 803 (2019)

41. Currie, G., Reynolds, J.: Vehicle and pedestrian safety at light rail stops in mixed traffic. Transp. Res. Rec. **2146**, 26–34 (2010)
42. Tubis, A., Rydlewski, M., Budzyński, M.: Safety assessment of tram stops. J. KONBiN **49** (2), 431–458 (2019)
43. Podciborski, T., Kopylov, O.: Proposal of method for the assessment of stops shelves in public transport in terms of the safety of their users (in Polish). Autobusy **6**, 403–405 (2016)
44. Solecka, K., Nosal Hoy, K., Deryło, A.: Assessment of transport interchanges for the needs of people with reduced mobility. Travel Behav. Soc. **21**, 48–56 (2020)
45. Ye, Z., Wang, C., Yu, Y., Shi, X., Wang, W.: Modeling level-of-safety for bus in China. Traffic Inj. Prev. **17**(6), 656–661 (2016)
46. Iseki, H., Taylor, B.D.: Style versus service? An analysis of user perceptions of transit stops and stations. J. Public Transp. **13**(3), 39–63 (2010)
47. Delbosc, A., Currie, G.: Modelling the causes and impacts of personal safety perceptions on public transport ridership. Transp. Policy **24**, 302–309 (2012)
48. dell'Olio, L., Ibeas, A., Cecin, P.: The quality of service desired by public transport users. Transp. Policy **18**, 217–227 (2011)
49. Soczówka, P., Żochowska, R.: Interactions between tram passengers and road vehicles at tram stops – a pilot study. Trans. Transp. Sci. **11**(2), 64–76 (2020)
50. Rydlewski, M., Skupień, E.: The influence of the type of traffic organization on the behaviour of pedestrians at tram tracks crossings. In: Macioszek, E., Kang, N., Sierpiński, G. (eds.) Nodes in Transport Networks – Research, Data Analysis and Modelling, LNiTI, pp. 154–170. Springer, Cham (2020)
51. Cheranchery, M.F., Bhattacharyya, K., Maitra, B., Boltze, M.: Assessing safety level of bus stops in the absence of crash data. TRB 2016 Annual Meeting, pp. 1–16 (2016)

Application of Genetic Algorithms for the Planning of Urban Rail Transportation System

Leonid A. Baranov[1], Valentina G. Sidorenko[1,2], Anthon I. Safronov[1], and Kyaw Min Aung[1(✉)]

[1] Russian University of Transport, Moscow, Russian Federation
[2] National Research University "Higher School of Economics", Moscow, Russian Federation

Abstract. The article introduces an integrated approach for the planning of urban rail transport system with the purpose of maintaining the high traffic density, ensuring the safety of trains, passenger comfortable service and properly use of resources. It explores the possibility of using genetic algorithms for all three types of planning the transportation process such as the planned train schedule, maintenance schedule and working schedule for locomotive crews. The main objective of the study is to improve an automated train planning system with the uniformity criterion by using a variety of resources under limited restrictions. The authors examined the possibility of genetic algorithm for the planning of urban rail transportation system by using various types of crossover, mutations and the influence of genetic algorithm parameters on the results of transportation planning process.

Keywords: Planning · Urban rail transport system · Genetic algorithm · Planned train schedule · Maintenance schedule · Working schedule for locomotive crews

1 Introduction

Maintaining the high traffic intensity, ensuring the safety of the urban rail transport system, observing the rules of comfortable passenger service and properly use of resources are based on an integrated approach of the transportation planning process [1, 2]. The planned train schedule is the result of transportation planning process. The passenger metro trains are closely related to the maintenance schedule of electric rolling stock and the working schedule of locomotive crews [3]. Many articles have been discussed the problem of urban transportation planning system by various methods [4–8]. The paper [9] examines the information system for automating the process of urban passenger transport system. The work of [10] describes the optimization model for working schedule of crews according to the interaction and restrictions of the European agreement. In the work of Zherbina [11], this problem is described as a classical assignment problem. The research of Sidorenko [12] discusses the formalization of the maintenance schedule by using the graph theory and Bellman's principle.

E. Macioszek and G. Sierpiński (Eds.): TSTP 2021, LNITI, pp. 21–39, 2022.
https://doi.org/10.1007/978-3-030-93370-8_2

The use of graph theory and Bellman's principle allows to get all possible sets of maintenance services and choose one which will correspond to the planned train schedule and minimal differ from the optimal according to the chosen criterion. But it takes a lot of time and the problem is when we need to choose the criterion of uniformity maintenance under limited resources. Therefore, the main purpose is to reduce the time of planning for the urban rail transportation system and to improve the automated planning system for the transportation process. In this case, the genetic algorithm is an effective tool for the optimization process [13–15].

The article examines the possibility of using genetic algorithms for all three tasks of planning the transportation process such as planned train schedule, maintenance schedule and working schedule for locomotive crews. The study intends to improve an automated system of train schedules with the uniformity criterion by using a variety of resources under limited restrictions. In this research, the mathematical models is based on the combinatorics, graph theory and genetic algorithms. The authors developed the adaption of the crossover and mutation of genetic algorithms according to the characteristics of the tasks. The research examined the possibility of genetic algorithm by using various types of crossover, mutations and the influence of genetic algorithm parameters on the results of transportation planning process.

The urban rail transport system includes metro, monorail systems, suburban railway and high-speed tram. Many papers have been discussed on the automation of planned train schedule for passenger metro which is important for the planning of urban rail transportation system and can be classified:

- general issues for the automation of planned train schedule for metro:
  - development of mathematical models for planned train schedule,
  - observation of business processes for the planned train schedule,
  - adjustments of interval time,
  - visualization,
  - planning for train maintenance,
  - planning for locomotive crews,
  - planning energy-optimal schedule,
- some other parts of planned train schedule:
  - traditional planned train schedule,
  - zone type schedule of metro and suburban railways,
  - circular line.

2 Approach

When we solve the optimization problems by using genetic algorithm, these following main steps must be carried out [16, 17]:

- initialization of the initial data for the operation of the algorithm:
  - importing the chromosome or individual that determined by the tasks,
  - importing data to calculate the fitness function,

  - setting the parameters of the algorithm (crossover, population size, the maximum number of iterations without improving the values of the fitness function, determining the accuracy of the value of fitness function, the maximum running time of the algorithm, etc.),
- creation of the first generation of the population,
- calculating the value of fitness function for each chromosome,
- the formation of a new generation, modeling of the evolutionary process:
  - individuals with the best fitness function values that are part of the current population,
  - individuals from the result of crossover of individuals selected as parents of the current population,
  - individuals from result of mutations of individuals selected as parents of the current population.
- checking the termination condition for the algorithm. If the termination is not occurred, go to step (c).

3 Methods for Solution

It is necessary to adapt the concepts of genetic algorithm for the problems of planning the transportation process. The results of this adaptation are shown in Table 1. The genotype is specified by a set of individual chromosomes when solving the problem of planned train schedule. Each of chromosome consists of genes and each of gene has its own set of alleles. The number of locus in each chromosome is equal to the number of places where the choice for planned train schedule takes place. The number of element in the set of alleles for each of locus is determined by the number of variants of planned train schedule at this stage. The description of the genotype is given in Table 2, and Tables 3, 4 and 5 contain the information of various types of chromosomes.

Table 1. The adaption of genetic algorithm for the planning of transportation process.

Terms of genetics	Similar technical term	Definition	Adaptation for the planned train schedule	Adaptation for the maintenance schedule	Adaptation for the working schedule of locomotive crews
Population, generation	Set of individuals (chromosomes)	A finite set of individuals at the evolution iteration	Set of train schedule variants at the evolution iteration	A set of options for maintenance schedule at the evolution iteration	The set of options for the working schedule of locomotive crews at the evolution iteration
Individual	Individual, specimen	Genotype or a single chromosome if the genotype consists only a single chromosome	An instance of planned train schedule which described by a set of chromosomes presented in Table 2	Coincides with the chromosome, since the individual is described by one chromosome	Coincides with the chromosome, since the individual is described by one chromosome

(continued)

Table 1. (*continued*)

Terms of genetics	Similar technical term	Definition	Adaptation for the planned train schedule	Adaptation for the maintenance schedule	Adaptation for the working schedule of locomotive crews
Chromosome	Chain, binary sequence, string, or code sequence	Ordered gene sequence	A structure the describes one of the transient processes of planned train schedule. Chromosome construction options are presented in Tables 3, 4 and 5	A structure that describes the option of maintenance schedule in which each inspection that must be performed is assigned a place and time of its conduct	A structure that describes the variant of the working schedule of locomotive crews in which each change of the locomotive crews is assigned to perform
Gene	Property, symbol, or detector	Element of the genotype in particular chromosome	Presented in the Tables 3, 4 and 5	A tuple that contains the information about place and time of a specific inspection	Locomotive crew which is assigned to perform a specific change for the working schedule of locomotive crews
Genotype	Structure	Set of individual chromosomes	A set of individual chromosomes, which specifies the method for the corresponding planned train schedule instance, presented in Table 2	Chromosome, a set of tuples which is ordered by the inspections, that contains the information about the place and time of a specific inspection	Chromosome, which is a set of locomotive crews ordered by the schedule for the implementation of changes in locomotive crews
Locus	Location	The location of a particular gene in the chromosome	Presented in the Tables 3, 4 and 5	One of the inspections which must be specified by one element of chromosome	One of the replacements for personal crew schedule which must be specified by one element of chromosome
Allele	One of the alternative states of a gene	The form of genes that occupy the same locus in chromosomes and determine the phenotypic differences of the same attribute	Presented in the Tables 3, 4 and 5	One of the elements of tuple, containing information about the place and time of a specific inspection which can be assigned to perform an inspection	One of the locomotive crews who can be assigned for one of the replacements for locomotive crew schedule

(*continued*)

Table 1. *(continued)*

Terms of genetics	Similar technical term	Definition	Adaptation for the planned train schedule	Adaptation for the maintenance schedule	Adaptation for the working schedule of locomotive crews
Fitness-function	The objective function for the individuals of population	The accuracy measurement or approximate measurement of a solution to a problem	Deviation time from the planned schedule on the main track or the number of night stay points which cannot assign night placement	Criterion of uniformity maintenance	Criterion of uniformity working time for locomotive crew
Mutation	Changing of a gene	Changing in the genotype that occurs under the influence of the external or internal environment	Changing in the genotype (presented in Table 2) which occurs under the influence of the external or internal environment	Changing one of the elements of tuple which contains the information of place and time of an inspection under the influence of the external or internal environment	Changing one of the locomotive crews who is assigned for the replacement of locomotive crew schedule under the influence of the external or internal environment
Crossover	Exchange of genetic material between chromosomes	The basic genetic operator which exchanges genetic material between individuals and simulates the process individuals	Exchange of genotype (presented in Table 2) as a result of the exchange of genetic material between individuals in a population	Exchange of tuple element value which contains the information of place and time of an inspection) as a result of the exchange of genetic material between individuals in a population	Exchange of locomotive crews who is assigned for the replacement of locomotive crew schedule as a result of the exchange of genetic material between individuals in a population
Evolution	Process	The process of population change until to the accomplishment of algorithm criterion	The process of variants change in the planned train schedule	The process of variants change in the train maintenance schedule	The process of variants change in the locomotive crew schedule

Table 2. Genotype.

Type of chromosome	Abbreviation	The number of chromosomes in the genotype
Entering trains before morning rush hour	$a_{tr^{m}+}$	The total number of the trains which included in the transition process, multiplied by two (for the two tracks are considered)
Exiting from the night stay points	a_m	1
Withdrawing trains after morning rush hour	$a_{tr^{m}-}$	The total number of the trains which included in the transition process, multiplied by two (for the two tracks are considered)
Entering trains before evening rush hour	$a_{tr^{e}+}$	The total number of the trains which included in the transition process, multiplied by two (for the two tracks are considered)
Withdraw trains after evening rush hour	$a_{tr^{e}-}$	The total number of the trains which included in the transition process, multiplied by two (for the two tracks are considered)
Leaving for the night stay point	a_e	1
Appointment for maintenance	a_{nepik}	1

The possible options for fitness functions of the corresponding tasks are the followings. When creating a planned train schedule, the main criteria is the uniformity [19, 20]:

- uniformity intervals by departure of trains from stations,
- uniformity arrangement of train entry or withdraw.

The comparison of planned train schedule options is carried out according to the following:

- number of route exchanges by the depot,
- parameters of adjusting slopes on station tracks of the line (number (total), average,
- maximum duration and other parameters of adjusting slopes,
- completion times for each of the main tracks,
- parameters of the zone type schedule or trains withdraw/entry at intermediate stations,
- parameters of train maintenance times at terminal and intermediate stations.

Table 3. Chromosome structure for transitional processes of entry (withdraw) of metro trains before (after) the morning (evening) rush hour $a_{tr^{m+}}$, $a_{tr^{m-}}$, $a_{tr^{e+}}$ and $a_{tr^{e-}}$.

Type of gene	Number of genes for the given type	The number of elements in a set of alleles
Options for entry (withdraw) by the 1st track	The number of uniform entry (withdraw) within the transitional processes by the 1st track	The number of options for the implementation of each uniform entry (withdraw) within the transitional processes by the 1st track, $x_{i,j}$ [18]: $x_{i,j} = \frac{M_{i,j}}{\mathrm{GCD}(M_{i,j}, M_{i+1,j})}$, $M_{i,j}$ – the number of trains that must be on the j-th track of the line at the beginning of the interval time with sequence number i, $M_{i+1,j}$ – the number of trains that must be on the j-th track of the line at the beginning of the next interval time, GCD – greatest common divisor
Options for entry (withdraw) by the 2nd track	The number of uniform entry (withdraw) within the transitional processes by the 2nd track	The number of options for the implementation of each uniform entry (withdraw) by the 2nd track which is calculated in the same way as for the 1st track
Routes for a specified train night placement	The number of train entry (withdraw) within the transitional processes	Number of routes that can be assigned for the corresponding line

The provision of small uniformity criteria values is achieved by the implemented algorithms. The end times of the first and second main tracks are used as a fitness function. It helps to compare the quality of options for the planned train schedule. We should pay attention to these indicators in order to be in time all the train movements along the metro line. This is necessary to provide the time as long as possible for night placement. The previous results have been shown that, despite the potentially large number of options for choosing the planned train schedule with the given initial data, the number of options for the successful schedule is still not enough. Therefore, another criterion for a successful planned train schedule is the number of night placement pointers where can be assigned the trains which failed to assign the night placement with the selected options. If the schedule is successful, there is no need such pointers. As a criterion for the rational planning of the maintenance services, the uniformity of the placement of services was introduced which is determined by one of those methods:

- as the sum of the squared deviations of maintenance start times for the candidate from the desired service start times,
- as the sum of the squares of the time intervals between services.

Table 4. Structure of chromosomes a_m and a_e.

Type of gene	Number of genes for the given type	The number of elements in a set of allele
Path in the tree diagram	1	Number of path in the tree diagram
Route to a given night stay point	Length of path in the tree diagram	Number of routes which can be assigned to the given night stay point

Table 5. Structure of chromosomes a_{nepik}

Type of gene	Number of genes for the given type	The number of elements in a set of alleles
Route to the inspection	The number of inspections which have been carried out by all technical inspection points	The number of routes which need to carry out an inspection during the current period of time at the given technical inspection points

In the case of limited resources, the value of the total exceed time between services over the admissible time is used as a criterion of solution for any initial data. For the locomotive crew schedule, the criterion is determined by the balance between the skills and the number of the personnel who performing the regular requirements. In our case, the criterion is the standard deviation of the working time of various locomotive crews during a period of time.

4 Determining the Size of the Primary Population

The process of adjusting the parameters of the genetic algorithm includes the population size, which determines how many individuals are present in each generation. With a large population size, the genetic algorithm searches the solutions more carefully and reducing the probability that only a local rather than a global minimum will be found. At the same time, a large population size leads to the fact that the algorithm will run slower. The size of the population can range from one to maximum number of options for constructing a planned train schedule, which is equal to the product of the number of options for implementing all stages of the planned train schedule.

With a population size equal to the maximum number of options for the planned train schedule, genetic algorithm degenerates into an exhaustive search algorithm, and population size 1 is a degenerate case (random choice of any value and declaring it optimal in a voluntarist manner). If the size of the primary population is determined by the maximum value of two quantities:

- the maximum value among the powers of the set of alleles of all genes of chromosomes described in Tables 2, 3, 4 and 5, except for genes associated with assigning a route to a string, delivering it to the night stay point or sending it for inspection. In this case, it guaranteed ensured the presence of all possible alleles for each of the genes,
- the maximum number of permutations of alleles between locus when assigning routes to threads, delivering them to night placement or when sending them for inspection. In this case, the calculation is calculation in this case is as shown below by calculating the size of the population when solving the problem of maintenance schedule.

Let the population size be N. Our objective is to find the number of individuals at which the chromosomes of an individual to get all the values of elements for the complete set by recombination. For this, the population size must be for each locus in the entire population of all possible allele. Let's calculate the probability that a random set of chromosomes will contain all possible allele values at the selected locus. The probability that any one of the locus will contain the same value in all chromosomes is equal to $\frac{1}{(N_a)^N}$. In a population, a complete set of alleles at one of the locus may be present in the case of $N \geq N_a$. In this case, in N_a chromosomes the values differ, and in the remaining ones they take arbitrary values. The number of realizations of filling the locus of the same name in the population, in which the chromosome will contain all possible values of the alleles, in the selected locus, is determined by the following expression:

$$\overline{C}_{N_a}^{N-N_a} = C_{N_a+N-N_a-1}^{N-N_a} = C_{N-1}^{N-N_a} = \frac{(N-1)!}{(N-N_a)!(N-1-N+N_a)!}$$
$$= \frac{(N-1)!}{(N-N_a)!(N_a-1)!} \quad (1)$$

where:

$\overline{C}_{N_a}^{N-N_a} = C_{N_a+N-N_a-1}^{N-N_a}$ – the number of combinations with repetitions of alleles N_a in the same locus of all remaining after filling with different alleles $N - N_a$ chromosome,

N_a – the number of elements in a set of allele.

In total, there are $\overline{C}_{N_a}^{N} = C_{N_a+N-1}^{N} = \frac{(N_a+N-1)!}{(N)!(N_a-1)!}$ options for filling the same locus in the population. Then the probability that a random set of chromosomes will contain all possible allele values at the selected locus is:

$$
\begin{aligned}
p_1 &= \frac{\text{Number of favorable events}}{\text{Number of all events}} = \frac{\frac{(N-1)!}{(N-N_a)!(N_a-1)!}}{\frac{(N_a+N-1)!}{(N)!(N_a-1)!}} \\
&= \frac{(N-1)!(N)!(N_a-1)!}{(N-N_a)!(N_a-1)!(N_a+N-1)!} = \frac{(N-1)!(N)!}{(N-N_a)!(N_a+N-1)!} \\
&= \frac{(N_a+N-1-N_a)!(N)!}{(N-N_a)!(N_a+N-1)!} = \frac{A_N^{N_a}}{A_{N_a+N-1}^{N_a}} = \frac{N\cdot(N-1)\ldots(N-N_a+1)}{(N_a+N-1)\cdot(N_a+N-2)\cdot\ldots\cdot N} \\
&= \frac{(N-1)\ldots(N-N_a+1)}{(N_a+N-1)\cdot(N_a+N-2)\ldots(N+1)}
\end{aligned} \tag{2}
$$

where:

$A_N^{N_a}, A_{N_a+N-1}^{N_a}$ – number of placements.

Accordingly, the probability that all N_c locus will contain the full set of alleles each is equal to:

$$
p_{N_a} = (p_1)^{N_c} = \left(\frac{A_N^{N_a}}{A_{N_a+N-1}^{N_a}}\right)^{N_c} \tag{3}
$$

The probability of the locus which will not contain the full set of alleles is:

$$
1 - p_{N_a} = 1 - \left(\frac{A_N^{N_a}}{A_{N_a+N-1}^{N_a}}\right)^{N_c} \tag{4}
$$

Figure 1a, 1b, and 1c shows the results of the calculation of the following probabilities for one of the lines of the Moscow metro according to formulas (2)–(4):

- the probability that a random set of chromosomes will contain all possible allele values at the selected locus,
- the probability that all N_c locus will contain the full set of alleles,
- the probability of the locus which will not contain the full set of alleles.

In this case, $N_a = 38$ and $N_c = 13$.

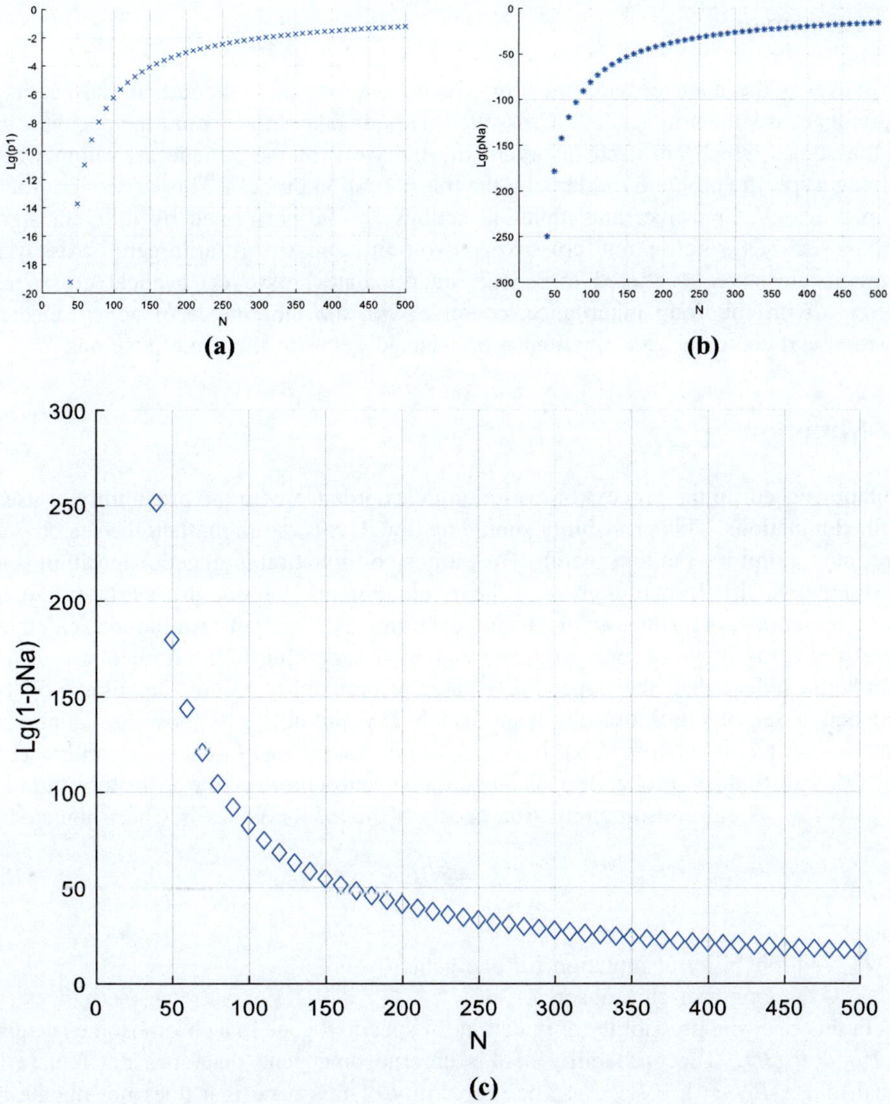

Fig. 1. a. The probability that a random set of chromosomes will contain all possible allele values at the selected locus, b. The probability that all N_c locus will contain the full set of alleles, c. The probability of the locus which will not contain the full set of alleles

5 Crossover

Crossover is the main genetic operator which exchange the genetic material between individuals as shown in Table 1. Crossover methods take into account the peculiarities of the alleles which are used in organizing the work of the genetic algorithm when solving a specific problem of planning the transportation process. This crossover differs from previous. Crossover and mutation actions can be performed by different algorithms such as single-point crossover, two-point crossover, arithmetic crossover, heuristic crossover, scattered crossover, intermediate crossover, cyclical crossover, coercive with subgroup inheritance, coercive with the inheritance of genes selected position and coercive with inheritance of selected genes to the same positions.

6 Mutation

Mutations occur in the process of evolution in accordance with the probability of user-defined mutations. This probability should be low. If it is too high, then the search will turn into a primitive random search. The purpose of mutation in genetic algorithm is to represent diversity. Mutation prevents the termination of evolution process after finding the local minima, or a situation in which the chromosome are very similar to each other, thereby slowing down or even stopped evolution. According to the limited size of the population is less than the upper boundary, the probability of any iterations will be obtained a set of chromosomes from which any possible chromosome cannot be obtained by recombination. This is why we used mutation in genetic algorithm [13, 21]. We can find the probability of mutation for one chromosome to restore the lost variants P_{mc}. A chromosome mutation occurs if one of the genes has been mutated:

$$P_{mc} = N_c P_{m1} \tag{5}$$

where:

P_{m1} – probability of mutation for one gene,

N_c – the length of chromosome.

In this case, the probability of mutation for specified gene in a chromosome is equal to $P_{m1} = P_{mc}/N_c$. The probability of this chromosome gene that does not mutate is equal to $1 - P_{m1} = 1 - P_{mc}/N_c$. The probability of this gene that does not mutate in any chromosome is equal to $(1 - P_{m1})^N = (1 - P_{mc}/N_c)^N$. Accordingly, the probability of a given locus that mutates in at least one of the chromosomes is equal to $1 - (1 - P_{m1})^N = 1-(1 - P_{mc}/N_c)^N$. For reliability, we need to exceed the probability of a given locus mutation in chromosomes (at least one of the chromosomes) is larger than the probability of the value absence in this locus of all chromosomes. If the number of mutation is bigger than the entire set of alleles, it already does not mean mutation. Therefore:

$$
\begin{aligned}
1-(1-p_{m1})^N &> 10(1-p_{N_a}) \\
\frac{1-(1-p_{m1})^N}{10} &> (1-p_{N_a}) \\
\frac{1-(1-p_{m1})^N}{10}-1 &> (-p_{N_a}) \\
-\frac{1-(1-p_{m1})^N}{10}+1 &< (p_{N_a}) \\
p_{N_a} &> 1-\frac{1-\left(1-\frac{p_{mc}}{N_c}\right)^N}{10}
\end{aligned}
\tag{6}
$$

If the population is large, then the mutation is meaningless:

$$
1-\left(1-\frac{p_{mc}}{N_c}\right)^N > 10\left(1-\left(\frac{A_N^{N_a}}{A_{N_a+N-1}^{N_a}}\right)^{N_c}\right)
\tag{7}
$$

Hence:

$$
\begin{aligned}
p_{mc} &> N_c\left(1-\left(1-10\left(1-\left(\frac{A_N^{N_a}}{A_{N_a+N-1}^{N_a}}\right)^{N_c}\right)\right)^{1/N}\right) \\
&, p_{mc} > N_c\left(1-(1-10(1-p_{N_a}))^{1/N}\right)
\end{aligned}
\tag{8}
$$

7 Results

In this research, the authors developed a decision support system for the planning of the transportation process by using the genetic algorithm. Figures 2, 3, 4 and 5 show the results of train maintenance services planning for one of the Moscow metro lines by using the decision support system.

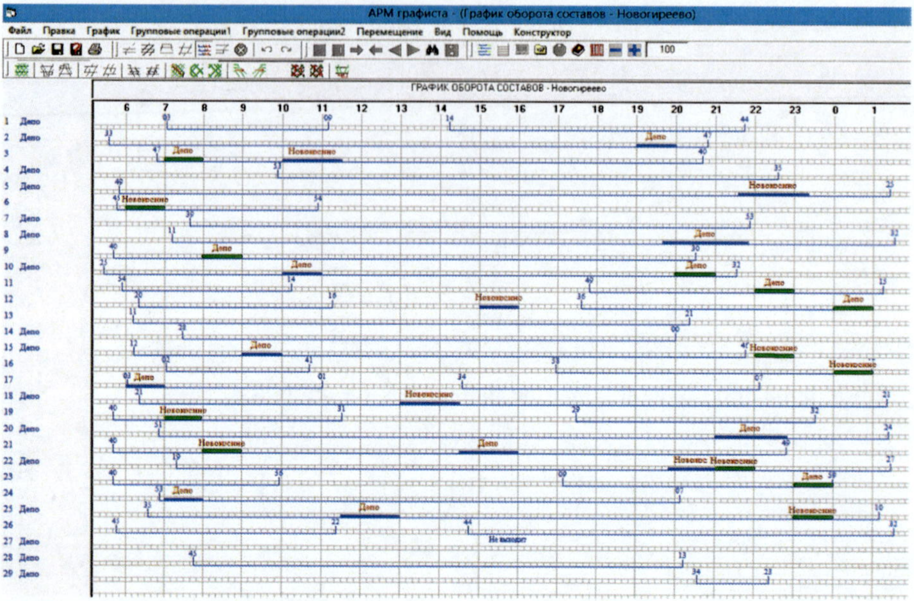

Fig. 2. The result of maintenance schedule while the resources are enough for the necessary condition of maintenance scheduling

Figure 2 shows the results of maintenance schedule in which the resources are enough for the maintenances of electric trains and fitness function of genetic algorithm reflects the uniformity of maintenances services. This figure allows to compare with the corresponding maintenance schedule which is using in Moscow metro. The blue rectangles are the real maintenances and the green rectangles are the proposed maintenances. Analysis of the result data shows the following:

- maintenances of the metro trains are carried out more often than the safety requirements. There are 18–23 maintenances on the schedule, but according to the safety requirements, there should be 13. This is showing that there are sufficient resources to carry out inspections and the number of trains on the line in non-rush hour is one third less than the maximum,
- according to the planned train schedule for metro, many routes are neither start running from the beginning moment of applying voltage to the contact rail nor until to the end of the day. Only one third of the routes start before 6 am or end later than 1 am.

In this regard, it makes sense to consider the situations of limited resources for maintenance schedule. Figure 3 shows the corresponding maintenance schedule in which the set of candidates that can be used for inspections has been decreased to a half.

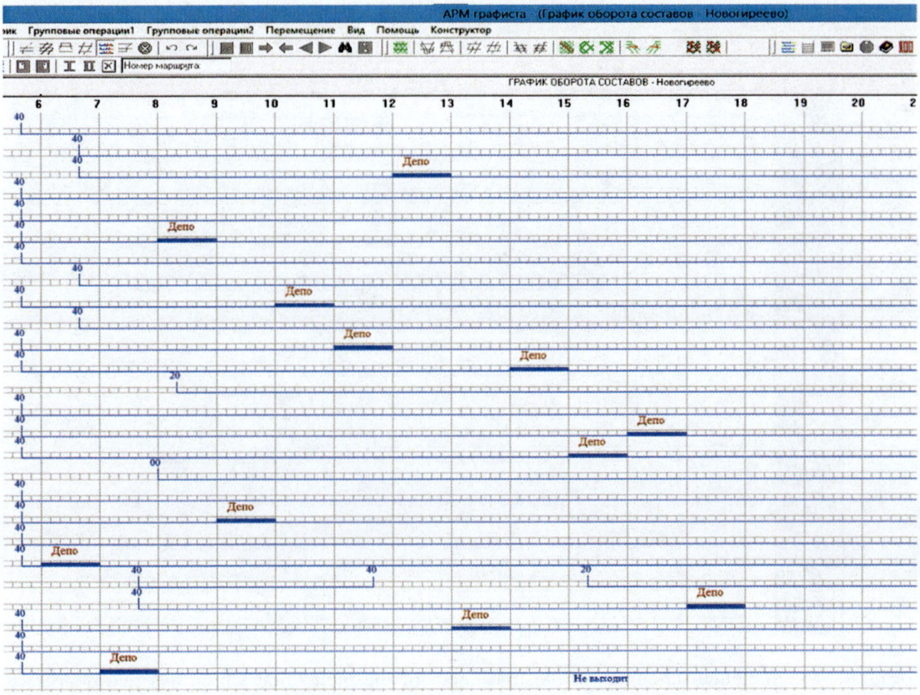

Fig. 3. The result of maintenance schedule while the resources are enough for the necessary condition of maintenance scheduling

In this case, the planned train schedule must reduce the running time interval between two inspections by early departure or late exit from the night stay point. The quadratic criterion was used for the solution and the "imaginary" inspection was on a chain of 10–14 routes. If sufficient resources are available, this chain should accommodate three inspections, not two.

The decision support system takes into account various features of the metro lines, for example, the presence of two depots and one linear inspection point on the line, which can carry out the train maintenance of both depots. The restrictions of maintenance schedule affect the formation of primary population. In this case, at first, for each of the depots, all the possibilities of inspections or the technical maintenance for the routes of this depot are distributed between the inspections of the same depot. After that, the remained inspections are determined. For their implementation, candidates related to the technical inspection places are appointed, who can serve all the depots of the line. The maintenance schedules for such conditions are presented in Figs. 4 and 5.

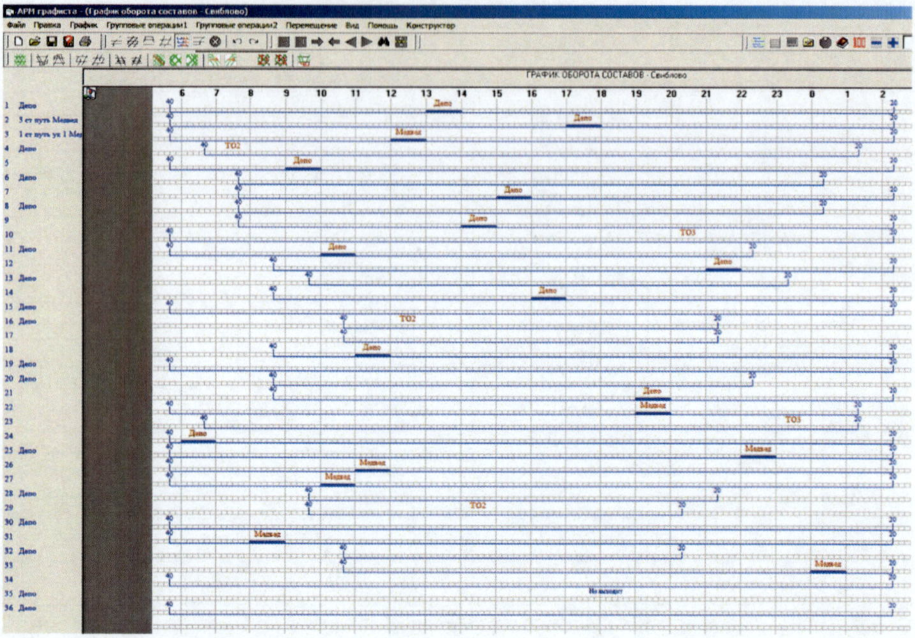

Fig. 4. Maintenance schedule for the first depot of the metro line which has two depots

Initially, we get the size of the population. Then we create a matrix based on the population to work in the program. Further, according to our code snippet, we understand which train is going to which depot, therefore:

- if we get 1, then the train goes to the Svivlovo depot,
- if we get 2, then the train goes to the Kaluzhskoye depot,
- if we get 3, then the train goes to the Medvegkovo depot.

Based on these calculations, we can obtain maintenance schedule information for the conditions of limited resources. This information contains the initial data of the operation for calculating the maintenance schedule of electric rolling stock.

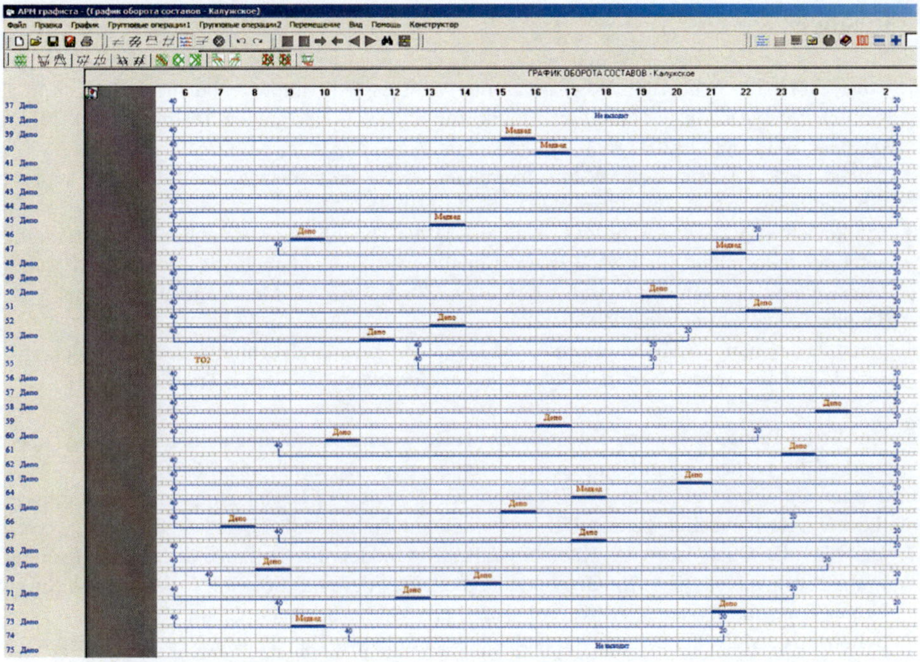

Fig. 5. Maintenance schedule for the Kaluzhskaya depot on the Kaluzhsko-Rizhskaya line of the Moscow metro

8 Conclusions

The article describes the adaptation of genetic algorithm for solving the problems of planning the transportation process such as the planned train schedule, the maintenance schedule of rolling stock and the working schedule for locomotive crews. The developed decision support system illustrates its capabilities for performing in adaptation by using various crossover types, mutations and different input parameters (such as population size, number of iteration, accuracy, etc.). It made possible to solve the problems of planning the movement of trains in the presence of limited resources by assuming the total excess time over the limited interval time as an additional criterion and the implementation of optimal maintenance schedules for the planning of urban rail transport system. The decision support system has been developed an interface for the data exchange of automated planned schedules of passenger trains for the Moscow metro [22], which is an example of the implementation of the centralized intelligent control system of the urban rail transport system.

Acknowledgments. The reported study was funded by RFBR, Sirius University of Science and Technology, JSC Russian Railways and Educational Fund "Talent and success", project number 20-37-51001.

References

1. Palte, M.H.A.: Smartfleet, How smart rail vehicles help improve business. In: 5th IET Conference on Railway Condition Monitoring and Non-Destructive Testing (RCM 2011), pp. 1–4. IET, Derby (2011)
2. Sauder, R.L., Westerman, W.M.: Computer aided train dispatching: decision support through optimization. Interfaces **6**(13), 24–37 (1983)
3. Markevich, A.V., Sidorenko, V.G., Kulagin, M.A.: The influence of the human factor on the safety of train traffic. In: XXVII International Scientific Conference on Problems of Safety Management of Complex Systems, pp. 265–270. Institute of Management Problems V.A. Trapeznikova, Moscow (2019)
4. Zhao, J., Chan, A.H.C., Burrow, M.P.N.: A genetic-algorithm-based approach for scheduling the renewal of railway track components. Rail Rapid Transit **223**, 533–541 (2009)
5. Ho, T.K., Tsang, C.W., Ip, K.H., Kwan, K.S.: Train service timetabling in railway open markets by particle swarm optimization. Expert Syst. Appl. **39**, 861–868 (2012)
6. Turner, C., Tiwari, A., Starr, A., Blacktop, K.: A review of key planning and scheduling in the rail industry in Europe and UK. J. Rail Rapid Transp. **230**(3), 984–998 (2016)
7. Davydov, B., Chebotarev, V., Kablucova, K.: The model of arrival time distribution forming in the dense vehicle flow. In: Grzegorz, S. (ed.) Integration as Solution for Advanced Smart Urban Transport Systems. AISC, vol. 1083, pp. 69–78. Springer, Cham (2019)
8. Schöbel, A., Maly, T.: Operational fault states in railways. Eur. Transp. Res. Rev. **4**(2), 107–113 (2012). https://doi.org/10.1007/s12544-011-0068-z
9. Nemtinov, V., Nemtinova, Y., Borisenko, A., Mokrozub, V.: Information support of decision making in urban passenger transport management. Transp. Probl. **12**(4), 83–90 (2017)
10. Oliskevych, M., Kovalyshyn, S., Magats, M., Shevchuk, V., Sukach, O.: The optimization of trucks fleet schedule in view of their interaction and restrictions of the European agreement of work of crews. Transp. Probl. **15**(2), 157–170 (2020)
11. Zherbina, A.I.: Calculation on the computer schedule turnover of trains metro. Transp. MIIT **612**, 100–105 (1978)
12. Sidorenko, V.G., Filipchenko, K.M.: The method of effective service planning with the application of graph theory. Informatization Educ. Sci. **4**(28), 123–132 (2015)
13. Mitchell, M.: An introduction to genetic algorithms. MIT Press, London (1999)
14. Tormos, P., Lova, A., Barber, F., Ingolotti, L., Abril, M., Salido, M.A.: A genetic algorithm for railway scheduling problems. In: Xhafa, F., Abraham, A. (eds.) Metaheuristics for Scheduling in Industrial and Manufacturing Applications. Studies in Computational Intelligence, vol. 128, pp. 255–276. Springer, Berlin, Heidelberg (2008)
15. Sheng, S., Dechan, Z., Xiaofei, X.: Genetic algorithm for the transportation problem with discontinuous piecewise linear cost function. Int. J. Comput. Sci. Network Secur. **6**(7(A)), 182–190 (2006)
16. Holland, J.H.: Adaptation in Natural and Artificial Systems. The MIT press, Cambridge, Massachusetts (1992)
17. Sidorenko, V.G., Aung, C.M., Alekseev, V.M., Rozenberg, E.N., Umanskii, V.I.: Planning electric-rolling-stock maintenance in conditions of limited resources. Russ. Electr. Eng. **88** (12), 839–841 (2017)
18. Safronov, A.I.: Automated construction of the planned schedule for the movement of passenger subway trains. Vestnik **3**, 99–104 (2012)

19. Seslavin, A.I.: Principles of uniformity in problems of control of passenger transport flows. Appl. Inf. **2**(20), 91–95 (2009)
20. Kontsevich, M.L.: Uniform arrangements. Kvant **7**, 51–59 (1985)
21. Goldenberg, D.E.: Genetic Algorithms in Search, Optimization and Machine Learning. Addison-Wesley, New York (1989)
22. O'Connor, R.V., Elger, P., Clarke, P.M.: Continuous software engineering – a microservices architecture perspective. J. Software Evol. Process **29**(11), 866 (2017)

Inland Navigation Point Infrastructure as a Source of Transport Planning Decisions

Emilia T. Skupień(✉)

Wroclaw University of Science and Technology, Wrocław, Poland
emilia.skupien@pwr.edu.pl

Abstract. The paper refers to transport planning decisions and their correlation to the infrastructure. The author aims to determine what factors related to transport infrastructure affect the transport planning process. The specific measure of the research is the area of inland navigation. The goal of the proposed research is modelling transport by inland waterways by influencing the infrastructure (its technical parameters, accessibility, information about it, etc.). The structure of the article includes a review of the present literature on infrastructure and the implementation of the transport process, inventory of the hydrotechnical infrastructure in Poland, the factors and actors creating the state of these elements, and conclusions. Based on the research method, it was estimated who and how can influence the infrastructure, but there is a lack of answer on how those actors and actions, influence the transport planning decisions. Answering these questions will be the next research step.

Keywords: Inland navigation · Inland waterways · Transport planning · Transport infrastructure · Inland waterway infrastructure · Transport planning decisions

1 Introduction

All transportation processes are carried out in the anthropo-techno-spherical system. Therefore, they are influenced by people (their behaviour and competencies), technical elements (infrastructure and vehicles), and the environment [1]. The article focuses on the infrastructure's influence on the transportation process.

Infrastructure influences cost, time, and conditions of transport. Parameters of linear infrastructure are directly related to the capacity of transported loads. Similarly, there is a correlation between economic stimulus and the presence of transport infrastructure [2]. However, the sources scarcely refer to point infrastructure in this context. The location of point infrastructure in railways and inland waterways is mostly decided and administered by central authorities. The point infrastructure related to road transport differs from other sectors -the location of some elements (gas stations, parking lots) is mainly determined by trading factors and decisions of investors. In each of mentioned systems, the point infrastructure is expected to influence the transport planning decisions, but it was not examined on which of parameters influences, and it was not measured how it influences.

E. Macioszek and G. Sierpiński (Eds.): TSTP 2021, LNITI, pp. 40–49, 2022.
https://doi.org/10.1007/978-3-030-93370-8_3

The condition of all the infrastructure (point and linear) of Polish inland waterways raises many reservations, it is unsatisfied and it is necessary to rebuild it [3]. Therefore the subject of the analyses in the article is to estimate the impact that the point infrastructure has on transport planning decisions and the transport process itself, with special consideration of inland navigation. So it could be taken into account before some of the investments will be made. The article aims to determine what factors related to transport infrastructure affect the transport planning process in inland navigation. Based on these results, a questionnaire will be developed for open interviews with researchers, users and administrators of shipping infrastructure to determine the manner of influence of individual factors. The global goal of the proposed research is the possibility of modelling transport by inland waterways by influencing the infrastructure (its technical parameters, accessibility, information about it, etc.).

The structure of the article includes a presentation of the adopted research methodology in point 2 with two sub-points: 2.1. methodology of literature review including the infrastructure and the implementation of the transport process, inventory of the hydrotechnical infrastructure in Poland, identification of point infrastructure and 2.2. methodology of open interviews focused on the factors and actors creating the transport process decisions. Next, in point 3, the results of the literature review and interviews are presented. On this basis, conclusions and assumptions needed to deepen the research by the survey are formulated and presented in point 4.

2 Research Method

To conduct the study, the research method was divided into two parts. The first part was a literature review in the area of transport infrastructure and the transport planning process, with particular emphasis on inland navigation. The second part presents the assumptions related to conducting interviews with experts related to the topic of inland waterway infrastructure.

2.1 Methodology of Literature Review

The topic of transport infrastructure is widely studied and described. For the literature review, the author used a multi-searcher Primo, a tool dedicated to reviewing the journals and publications databases for Wroclaw University of Science and Technology employees. The database was searched in the following steps. The first step gave over 53 000 responses to entry 'transport infrastructure', published last year. The second step narrow down the search by taking into account only titles of publications including the phrase, there were almost 500 responses last year, and almost 2 000 last five years. Nevertheless, the third step was adding the record 'planning' to the search, and that gave only 20 responses last year and over 80 last five years. Almost all of the answers are connected to the planning of infrastructure, not its connection to the transportation process. The fourth step searched for 'transport planning, which gave over 100 responses last year and almost 500 last five years. These were only studies devoted to infrastructure as a whole, not individual parts of it. Combining search terms 'infrastructure' and 'waterways' give only 70 responses from the last ten years. The summarise of the searched terms is shown in Table 1.

Table 1. The terms search in literature review and the responds.

Searched term	Period	Area	No of responds
Transport infrastructure	Last year	All the publication	>53 000
Transport infrastructure	Last year	Title	<500
Transport infrastructure	Last five year	Title	<2000
Transport infrastructure + planning	Last year	Title	20
Transport infrastructure + planning	Last five year	Title	80
Transport planning	Last year	Title	100
Transport planning	Last five year	Title	500
Infrastructure + waterways	Last year	Title	70

2.2 Methodology of Open Interviews

To determine the factors related to transport infrastructure affecting the transport planning process in inland navigation literature review was made. But as it was mentioned, not many authors combine parameters of point infrastructure with transport planning decisions. Therefore it was decided to define the factors based on open interviews with representatives of the world of science, practice and administration. Because of the distance between the researchers and the experts, and because of the time of pandemic – interviews were conducted via telephone and were recorded – after summarizing the answers, the records were deleted. The interview took place in March 2021. In total: two scientists, two inland navigation captains and two representatives of authorities were intervened. During the interview, each expert was asked about their professional opinions about the influence and the correlation between inland waterway infrastructure and the transport planning process. The conversations were undetermined, but each person was asked six basic questions:

- who influences inland waterway infrastructure?
- what are they influencing on?
- how are they influencing the infrastructure?
- how the infrastructure influences transport planning decisions?
- what transport planning decisions can be influenced?
- what is the correlation between infrastructure and planning decisions?

Each interview lasted at least about 1 h and gave a lot of information and suggestions. More answers were similar, and there were no contradictory statements. The results of open interviews answered the first three basic questions, but there was no clear answer to the other three. Therefore, on basis of the research, appeared a need to prepare a survey or questioner to be asked a wider group of stakeholders.

3 Results

3.1 Literature Review

Among last year publications on transport infrastructure problems, there is an approach to transport infrastructure resilience for disaster [4], so the technical parameters of objects. A similar area is described in [5], where the authors analyse reliability forecasting methods of the transport infrastructure technical safety. But most of the last five years papers refer to transport infrastructure in the context of investment efficiency and distribution. E.g. [6] refers to land management in the surroundings of the airport, taking into account the impact of transport accessibility and the growth of the economic development of the neighbouring region. And [7] proposes a research approach to infrastructure investments project outcomes and its uncertainty analysis method, focussing on the early stages of the project lifecycle.

Transport infrastructure problems are not only considered by civil or traffic engineer. One interesting approach is shown in [8] and [9]. The authors consider how are addressed social issues when planning the transportation infrastructure. There are opinions, that nowadays, transport infrastructure is planned based on the conviction that physical mobility has only advantages. But some publications [10] stand that instead of trying to meet the future mobility demand by constantly increasing the amount of infrastructure, alternatives that can support a sustainable future should be considered. Also, [11] and [9] undertakes similar aspects, pointing out that planning infrastructure should take into account environmental aspects, not only in the context of specific infrastructure facilities but entire network creation plans.

The short review shows that the transport infrastructure is an important measure, but not all of its aspects are being studied. Especially correlation between the infrastructure and the transport planning decision is missed.

Combining search terms ‘infrastructure’ and ‘waterways’ give only 70 responses from the last ten years. Part of them was devoted to investments in infrastructure and optimization of investments value [12–14]. However, searching for terms ‘inland navigation’ and ‘planning’ gives in return papers slightly connected to the topic of this paper. [15] describes the connection between fairway conditions, the ship’s propulsion system, and the captain’s behaviour with the attainable speed and fuel consumption of inland ships. Authors of [16–19] present algorithms for route planning in inland waters. However [16] takes into account obstacle on waterways (e.g. fishnets) rather than infrastructure. The whole [20] is devoted to inland water-way transportation risk assessment. [17, 18] to not refer to the influence of infrastructure, but show models in an existing environment without further consideration. The authors of [21] determine the accessibility and availability of inland waterway but do not connect it with transport planning decisions.

Preparing for the interviews based on the literature review, it is important to notice also the law regulations connected with inland waterways and inland navigation in Poland. Most important are: The Act of 20 July 2017 – Water Law [22], The Act of December 21, 2000, on inland navigation [23], Regulation of the Minister of Infrastructure of 28 April 2003 on shipping regulations on inland waterways [24] and Regulation of the Council of Ministers of 9 October 2020 on the classification of inland waterways [25]. Those documents, in terms of inland navigation infrastructure, regulate the parameters and ongoing maintenance of inland waterways, the time of navigation season, parameters and ongoing maintenance of hydrotechnical infrastructure (e.g. locks, weirs, culverts, dams), some aspects of ports and marinas and navigation signs – its presence and meaning.

Based on the literature review it must be stated, that in the analysed literature, there is not defined clear and common connections between the inland waterway transportation infrastructure, and transport planning decisions.

3.2 Open Interview

Summarising the open review, except for the direct answers to asked questions, some general conclusions were made. The most important function of transport infrastructure is to enable the transportation process, so the correlation between the infrastructure and the process is unquestionable. There is a lack of recent publication in this area, probably because of the basic matter of that issue. The transport needs determine the beginning and end of the route. The roles of infrastructure in traffic flow are shown in Fig. 1.

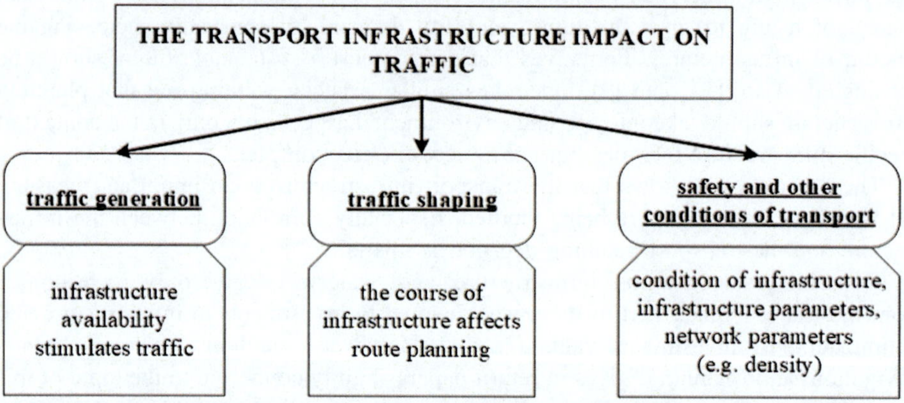

Fig. 1. Roles of infrastructure in traffic flow

Transport infrastructure also affects parameters of the transportation process and thus transport planning decisions. The presence of infrastructure gives the possibility to carry out the transport. Its availability (e.g. ship locks in Poland operate 16 h a day) and the network density influence the time of transport. Also, some technical conditions influence the route (e.g. height of bridges – important in the transport of over-size goods) or the speed (e.g. bad technical conditions of a surface). There are also some restrictions connected to transportation parameters: speed limits – influencing time, axle load limits – influencing capacity, and fees – influencing costs. Summarising it can be stated that transport infrastructure influence the time, costs and conditions of the transportation process. Also, there is a matter of branch of transport. If there is more than one available, then the infrastructure, and therefore parameters of transportation, plays an important role.

It is important to notice the difference between transport infrastructure in different transport branches. Inland waterways can be compared to railways, but they are much different from roadways. Railways and waterways and their infrastructure are mostly administrated by authorities and are a subject of centralized plans. Roadways net is denser and easier to access, so there are more commercial units nearby (e.g. petrol stations, restaurants, motels, intermodal hubs, private in-house roads). Even though, the infrastructure of inland waterways can affect the transport planning decisions. The first planning factor is: if inland navigation will be chosen to conduct the transportation at all. And that is the most common effect in Polish reality.

As it was mentioned before – the waterway exists only in specific places, their net is not dense and accessible nearby every transportation generation point. Bypassing on some sections of inland waterways is hardly possible and if some section is closed (e.g. due to lock renovation) it is almost impossible to find detour or alternative way using the same mean of transport. Therefore, more than in other branches of transport, inland navigation is determined by infrastructure. It does not however mean, that the influence cannot be measured and modelled.

The current state of Polish inland waterway infrastructure is assed as poor [e.g. 26, 27, 28]. Parameters of most Polish waterways are insufficient for international class IV of inland waterways [28, 29]. Only 5,5% of Polish waterways meet international criteria. The waterways network is not connected in one system but is a set of separate routes. However, in 2019, the national waterway network covered 3,722 km. Compared to other EU countries, this length is considerable. Also, the favourable natural conditions for the development of inland navigation are proven by a high network density index. In Poland, there is 11.7 km of navigable roads per 1000 km^2, and in the EU-28 on average 9.4 km/1000 km^2. On the other hand, out of 82 navigation locks located on inland waterways of particular importance for transport, only 5 have parameters suitable for at least class IV. Also, 123 bridges did not meet the minimum parameters defined for a waterway of international importance [30].

Moreover, it is important, that inland waterway infrastructure fulfil also other than transportation functions, which are shown in Fig. 2.

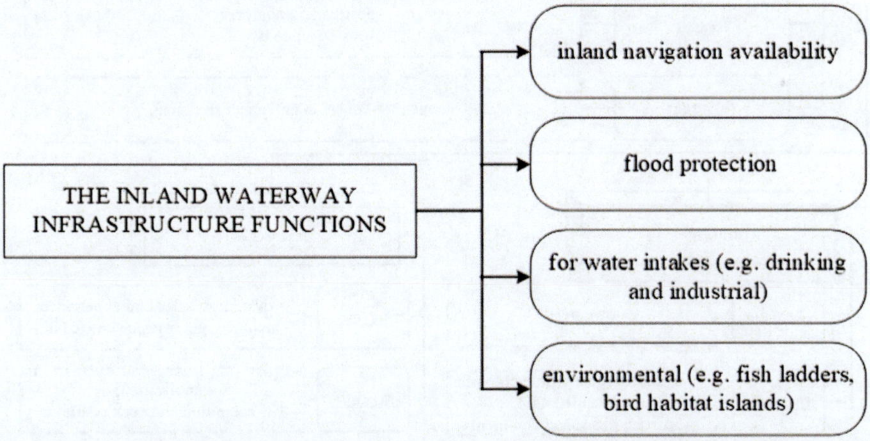

Fig. 2. Functions of inland waterway infrastructure

Due to multiple functions of inland waterways and the fact that Poland in 2016 joined European Agreement on Main Inland Waterways of International Importance (AGN) [31], which obligates Poland to improve the parameters of inland waterways, it is expected that soon the Polish hydrotechnical infrastructure will be rebuilt and some new objects will appear. That in consequence allows influencing future transport planning decisions, but it is necessary to assess the level and specific subject of influence. To assess who and how can influence the inland waterway infrastructure, the result of the research method were summarised. Most frequent answers to basic questions were collected and presented in Fig. 3.

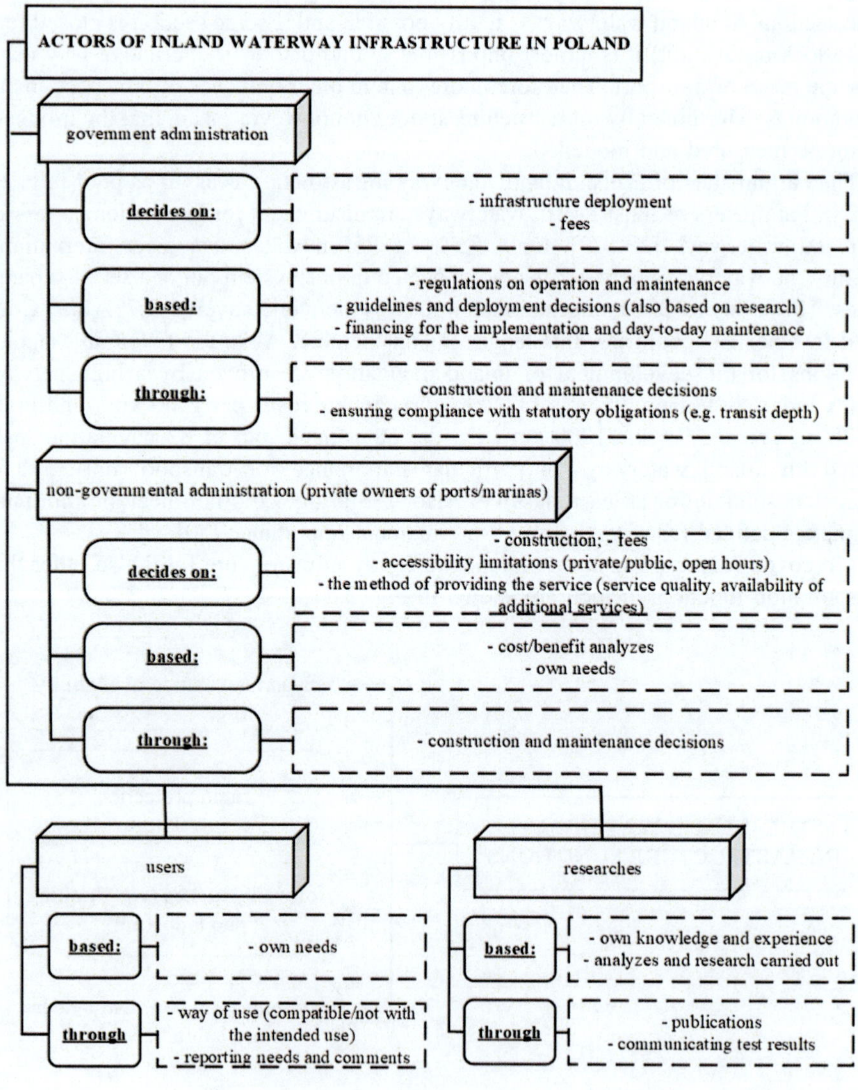

Fig. 3. The results of interviews research method

As it can be seen, experts determine four main groups of actors in case of influencing inland waterway infrastructure. These are government administration, non-governmental administration (private owners of e.g. ports/marinas), users of infrastructure and researchers. Both groups of administration have their influence via decisions on infrastructure location and parameters. Users demonstrate their influence by the way of using the infrastructure and potential needs reporting. Researchers may influence by the publications of their researches.

4 Conclusions

The data collected in the literature review and interviews did not fulfil the author's goal of the research. They do not answer what is the influence of inland waterway point infrastructure on transport planning decisions. Based on the research method, it was estimated who and how can influence the infrastructure, but there is a lack of answer on how those actors and actions, influence the transport planning decisions. As the influence is of significant interest to the author, the next steps were specified.

It was decided that the best way to establish the influence is to survey with closed questions a group of at least: 5 researchers, 5 representatives of authorities and 10 inland waterways transportation users. To get better results (and be able to take advantage of foreign good practices), at least half of the experts should base their response on foreign countries experience. Specifying, the questions should be constructed as follow: *How do you rate the influence on transport planning decision of ... ?* and concern specific aspects of the inland waterway infrastructure. E.g. cost of service, working hours, the density of occurrence, visibility of signs, the correctness of the information on signs, service time at locks, quality of service, additional services (availability of drinking water, possibility of leaving waste)… The list is still open.

Answers should include a five-point scale of influence, determining the opinion of an expert on how much some aspects connected with infrastructure, influence the transport planning decision. The question for users of inland waterways and shipper should be connected to the question of rating this parameter of the waterway, the expert knows about (with specifying the waterway). The survey should include questions of the role in inland waterway transportation (researcher/user/administrator) and the origin of experts experience (Polish/foreign/mixed). All of the questions, even if closed, should have an option to add some comments. In addition, asking the same questions to different groups of interest will help to assess the relationship between actions and regulations intention and reception. The survey may be distributed online to specific experts to avoid random opinions. It is also advisable to conduct an in-depth interview with a part of the experts to get a wider view of the theme.

The fact that inland navigation is currently rarely used to transport goods in Poland, and that the inland waterways in Poland are to be rebuilt shortly, the conducted research, also with the results of its planned continuation, may help with getting the increase of use of this efficient and ecological mean of transport.

References

1. Smalko, Z., Nowakowski, T., Tubis, A.: Zarys niezawodnościowej teorii zagrożeń. Navigator, no. **25**, ISSN 1425-0993 (2020)
2. OECD: Transport bridging divides. In: OECD Urban Studies. OECD Publishing, Paris, (2020). https://doi.org/10.1787/55ae1fd8-en
3. Resolution of the Council of Ministers of June 14, 2016 on the adoption of Assumptions for the development plans of inland waterways in Poland for 2016–2020 with 2030 perspective. http://mgm.gov.pl/wp-content/uploads/2017/11/assumptions-for-the-development-plans-of-inland-waterways-in-poland-for-2016-2020-with-2030-perspective.pdf (2016)
4. Salini, P.N., Saran, M.S., Sabitha, N.M., Sreeja, S.S.: Planning for resilient transport infrastructure for disaster prone area – case study of Munnar town. IOP Conf. Ser. Mater. Sci. Eng. **1114**(1), 012019 (2021). https://doi.org/10.1088/1757-899X/1114/1/012019
5. Tsarkova, E., Belyaev, A., Churakov, D., Andreeva, E.: Reliability forecasting for optimal planning of measures for maintenance of security systems of transport infrastructure facilities. IOP Conf. Ser. Mater. Sci. Eng. **918**, 012090 (2020). https://doi.org/10.1088/1757-899X/1114/1/012019
6. Dobryak, D.S., Novakovska, I.O., Nikolaiev, K.D., Skrypnyk, L.R.: Methods of imposing restrictions on land management in planning of an airport's transport infrastructure. IOP Conf. Ser. Mater. Sci. Eng. **708**(1), 012003 (2019). https://doi.org/10.1088/1757-899X/708/1/012003
7. O'Brien, P., Pike, A., Tomaney, J.: Governing the 'ungovernable'? Financialisation and the governance of transport infrastructure in the London 'global city-region.' Prog. Plann. **132**, 100422 (2019)
8. Antonsonab, H., Levincd, L.: A crack in the Swedish welfare façade? A review of assessing social impacts in transport infrastructure planning. Prog. Plann. **138**, 100428 (2020)
9. Mottee, L.K., Arts, J., Vanclay, F., Miller, F., Howitt, R.: Reflecting on how social impacts are considered in transport infrastructure project planning: looking beyond the claimed success of Sydney's South West rail link. Urban Policy Res. **38**(3), 185–198 (2020)
10. Givoni, M., Perl, A.: Rethinking transport infrastructure planning to extend its value over time. J. Plan. Educ. Res. **40**(1), 82–89 (2020)
11. Eckersten, S., Balfors, B., Gunnarsson-Östling, U.: Challenges and opportunities in early stage planning of transport infrastructure projects: environmental aspects in the strategic choice of measures approach. Sustainability **13**(3), 1–18 (2021)
12. Boudhoum, O.: Value Focused Inland Waterway Infrastructure Investment Decisions. ProQuest Dissertations Publishing, https://scholarworks.uark.edu/cgi/viewcontent.cgi?article=2295&context=etd (2015)
13. Bracarense, L., Brandão, R., Yamashita, Y., Aragão, J.: Feasibility study for waterway infrastructure: international overview and methodological recommendations. Transp. Res. Procedia **18**, 305–311 (2016)
14. Kielhauser, C., Martani, C., Adey, B.: Development of intervention programs for inland waterway networks using genetic algorithms. Struct. Infrastruct. Eng. **14**(5), 550–564 (2018)
15. Hekkenberg, R.G., van Dorsser, C., Schweighofer, J.: Modelling sailing time and cost for inland waterway transport. Eur. J. Transp. Infrastruct. Res. **17**, 508–529 (2017). https://doi.org/10.18757/ejtir.2017.17.4.3212
16. Kazimierski, W., Sawczak, A., Wawrzyniak, N.: Analysis of graph searching algorithms for route planning in inland navigation. In: TransNav, vol. 9, no. 2, pp. 281–286. Maritime University, Gdynia (2015)

17. Ivančić, P., Kasum, J., Pavić, I.: Proposed guidelines on developing the optimisation model for passage planning in inland waterways navigation. Pomorstvo **27**(2), 343–350 (2013)
18. Dubovan, L., Ghiba, M.: Technique of Using the "voyage planning" concept for the logistic integration of inland waterway transport. Adv. Eng. Forum **27**, 237–242 (2018)
19. Skupień, E., Hejducki, Z.: Modeling transport process using time couplings methods. Contemporary complex systems and their dependability. In: 13th International Conference on Dependability and Complex Systems DepCoS-RELCOMEX, pp. 430–439. Wroclaw University of Science and Technology, Brunów (2018)
20. Skupień, E., Tubis, A.: The use of linguistic variables and the FMEA analysis in risk assessment in inland navigation. In: TransNav, The International Journal on Marine Navigation and Safety of Sea Transportation, vol. 12, no. 1, pp. 143–148. Gdynia Maritime University, Gdynia (2018)
21. Bačkalić, T., Maslarić, M., Skupień E.: Analysis of navigation accessibility and fairway availability: a case study the middle Danube river. In: 10th International Scientific Conference, VII International Symposium of Young Researcher, pp. 53–62. Transport Problems, Katowice (2018)
22. Act of 20 July 2017 New Water Law (Journal of Laws 2017, item 1566). http://isap.sejm.gov.pl/isap.nsf/DocDetails.xsp?id=WDU20170001566 (2017)
23. Act of December 21, 2000 on inland waterway Navigation (Journal of Laws of 2000, No. 2020 item 1863). https://isap.sejm.gov.pl/isap.nsf/DocDetails.xsp?id=WDU20010050043 (2000)
24. Regulation of the Minister of Infrastructure of 28 April 2003 on shipping rules on inland waterways (Journal of Laws of 2003, No. 2003 No. 212 item 2072). http://isap.sejm.gov.pl/isap.nsf/DocDetails.xsp?id=WDU20032122072 (2003)
25. Regulation of the Council of Ministers of October 9, 2020 amending the ordinance on the classification of inland waterways (Journal of Laws No. 2020 item 1898). https://isap.sejm.gov.pl/isap.nsf/DocDetails.xsp?id=WDU20200001898 (2020)
26. Miklinska, J.: (Planned) Public private partnership projects for the development of Polish inland waterways. Economic and social development. In: 39th International Scientific Conference on Economic and Social Development – Sustainability from an Economic and Social Perspective, pp. 351–360. Book of Proceedings (2019)
27. Osowska, A., Skupień, E.: Transport possibilities at the section of lower vistula. J. KONBiN **43**(1), 223–240 (2017)
28. Rudnik, S.: Ratification by Poland of the European Agreement on major inland waterways of international importance (AGN). Transp. Overview Przegląd Komunikacyjny **2017**(10), 54–60 (2017)
29. Inland waterways transport in Poland in 2019. Statistical information. Szczecin, https://stat.gov.pl/obszary-tematyczne/transport-i-lacznosc/transport/transport-wodny-srodladowy-w-polsce-w-2019-roku,4,10.html (2019)
30. Inland waterways transport in Poland in 2018 and 2019, Statistical information. Warszawa, Szczecin, https://stat.gov.pl/en/topics/transport-and-communications/transport/inland-waterways-transport-in-poland-in-2018-2019,2,4.html, (2020)
31. European Agreement on Main Inland Waterways of International Importance (AGN), Geneva (19 January 1996), https://digitallibrary.un.org/record/582953 (1996)

Modelling and Optimization as Support in the Management of Transport Processes

Evaluation of BEV and FCHEV Electric Vehicles in the Creation of a Sustainable Transport System

Hubert Rzędowski[1](✉) and Ewelina Sendek-Matysiak[2]

[1] Department of Mechatronics and Mechanical Engineering, Kielce University of Technology, Kielce, Poland
hubertrz95@interia.pl

[2] Faculty Management of Modeling and Computer, Kielce University of Technology, Kielce, Poland
esendek@tu.kielce.pl

Abstract. The paradigm of sustainable development – equal treatment of economic, social, environmental and cultural spheres is the basis of development strategies and current economic policy in all developed and developing countries of the world. Currently, the aspect of ecological friendliness, i.e. the need to reduce the negative impact on the environment, is the main factor leading to the development of technology and organization in all areas of the economy. Meanwhile, according to studies, among the anthropogenic activities, the largest adverse impact on the environment is caused by transport activities, and in the transport sector itself. The largest source responsible for greenhouse gas emissions is road transport. Therefore, the motorization sector is constantly looking for new solutions to reduce its negative effects. One of them is the wide-scale dissemination of vehicles with electric engines. Currently, battery-powered vehicles are being produced and used on an increasingly large scale. There are also new solutions such as electric vehicles with fuel cells. The paper indicates which type of electric vehicles, battery-powered or fuel-cell-powered, is a better solution for shaping a transport system free of pollution and noise under present conditions. A comparative analysis of battery- and cell-powered vehicles was performed. It was observed that battery powered vehicles are currently a better option than cell powered vehicles. The biggest barrier to the development of cell-powered vehicles is the acquisition of hydrogen and construction of the infrastructure.

Keywords: Electric vehicles · Cell powered vehicles · Sustainability · Electromobility

1 Introduction

Since the period of industrialization, natural resources, especially raw materials, were intensively exploited, which had a negative impact on the environment (the side effect of the management process was and still is the emission of harmful gases, air and water pollution, etc.). Parallel to the process of industrialization and population growth,

E. Macioszek and G. Sierpiński (Eds.): TSTP 2021, LNITI, pp. 53–68, 2022.
https://doi.org/10.1007/978-3-030-93370-8_4

agricultural production was intensified, the development of which over time became an increasing threat to ecosystems through, among others, soil acidification, pollution of waters with artificial fertilizers. Consequently, this led to the extinction of some species of flora and fauna. The intensification of production processes was supposed, according to the belief and theory of economics, to ensure an increase in wealth. The accepted assumption that social development is based on the pursuit of material needs was revised over time, considering it incomplete [1]. The need began to be perceived to take into account in economic activity, in addition to the economic dimension, the role of social and environmental factors, the purpose of which would be to meet the needs of not only the current generation, but also future ones. Such approach corresponds to the concept of sustainable development.

From the beginning, the European Union countries actively supported activities aimed at establishing and then consolidating the concept of sustainable development.

E. Mazur-Wierzbicka in [2] points out that for the European Union, sustainable development is not a separate quality functioning in isolation from other areas of its activity. The Union accepts sustainable development as a universal direction of actions, which is to be present in all policies, including transport policy. Therefore, the goal of the European Union is to create such a transportation system that will respect the principles of sustainable development, i.e., a system that aims to ensure the availability of transportation services and the mobility of all residents in a safe and environmentally friendly manner, contributing to the economic development and increasing the welfare of society [3–6].

Transport, next to the energy sector, has been for many years the main source of greenhouse gas (CGH) emissions in the European Union (Fig. 1) – in 2018 it accounted for 25% of their emissions (Fig. 2), and in the transport sector itself the largest source responsible for greenhouse gas emissions is road transport [7].

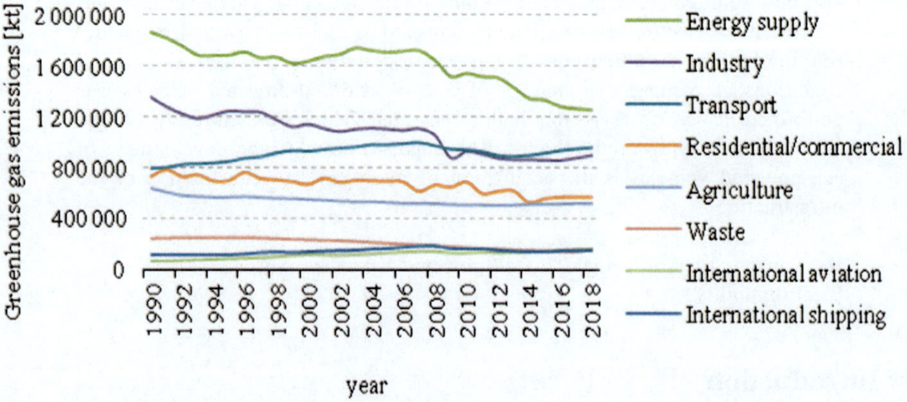

Fig. 1. Vehicles in use Greenhouse gas emissions by sector in the European Union from 1990 to 2018 (Source: Own elaboration based on[8])

Therefore, for many years now, numerous measures have been taken to reduce the adverse environmental impact of vehicles, such as the construction of drive systems that reduce pollutant emissions, the use of natural resources, and the amount of waste by appropriate recycling of end-of-life vehicles as well as the reduction of noise emission. However, the ever-increasing number of vehicles on the road (Fig. 3) means that this is still significant. The European Environment Agency predicts that in 2050 the demand for mobility in the EU will increase by about 60%.

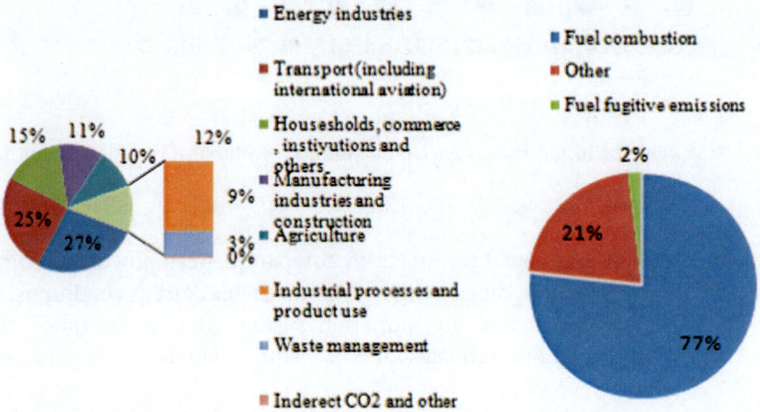

Fig. 2. Greenhouse gas emissions by Intergovernmental Panel on Climate Change (IPCC) source sector, EU-28, 2017 (Source: Own elaboration based on [9])

Therefore, the continuous growth of demand for transport requires taking further steps to limit its negative effects. In terms of suprastructure, the European Commission has indicated that the solution to the problems generated by transport activities, apart from improving vehicles with combustion engines (increasing their efficiency, reducing exhaust emissions by, among other things, using alternative fuels such as liquefied natural gas (LNG), compressed natural gas (CNG), and the second-generation biofuels), is the radical increase in sales of vehicles with alternative sources of propulsion, including those with electric drives [11–13]. The scale of changes to be introduced to the transport sector in the European Union in the coming years is best illustrated by one of the assumptions of the White Paper of March 28, 2011 on Roadmap to a Single European Transport Area – Towards a competitive and resource-efficient transport system – halving the number of conventionally-fuelled vehicles and their total elimination from urban traffic by 2050 [14]. According to [15, 16], by 2035 only fully electric vehicles will be sold in the EU.

This goal, which from today's point of view seems unrealistic because of many economic, technological, and social barriers, may be attained within the next few decades, taking into account continuous development and technological progress [17]. At present, the electric automotive market is dominated by BEVs (Battery Electric Vehicles); however, it has been pointed out in [18] that such vehicles are only a temporary solution and will be replaced in the future by FCHEVs (Fuel Cell Hybrid Electric

Vehicles). On the other hand, H.P. Lenz is of a different opinion, claiming that hydrogen vehicles will not be popular due to higher infrastructure construction costs [19].

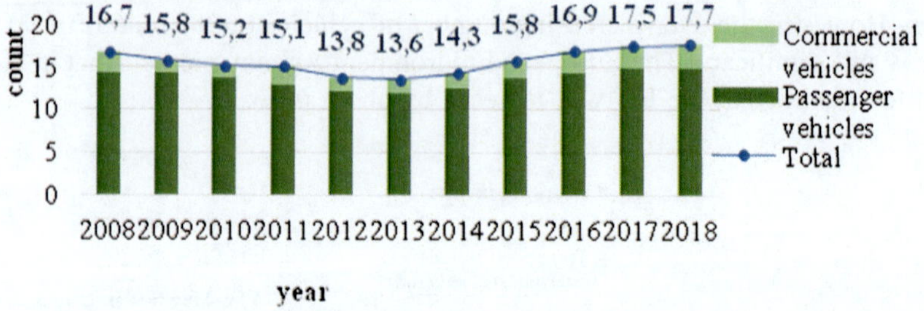

Fig. 3. Sales of vehicles in the European Union [million] (Source: Own elaboration based on [10])

In this paper, an attempt has been made to compare battery-powered vehicles and hydrogen vehicles in order to indicate which of them, under current conditions, is more effective in creating a sustainable transportation system. For this purpose, the construction, energy harvesting and infrastructure of electric vehicles have been analyzed.

2 Electric Vehicles Market

Within the European Union, in 2020, the number of BEVs was 11 25485 and the number of hydrogen vehicles was 1841 [20], while in Poland, 6556 passenger vehicles powered by battery electricity and 0 with fuel cells were registered at that time [20]. In April 2021, 23 834 BEVs were registered in Poland. Compared to 2020, the increase in newly registered vehicles was 137%. In 2021, 6 fuel cell vehicles were also registered [21]. Figure 4 presents the number of battery- and hydrogen-powered vehicles in the European Union and Poland in 2020.

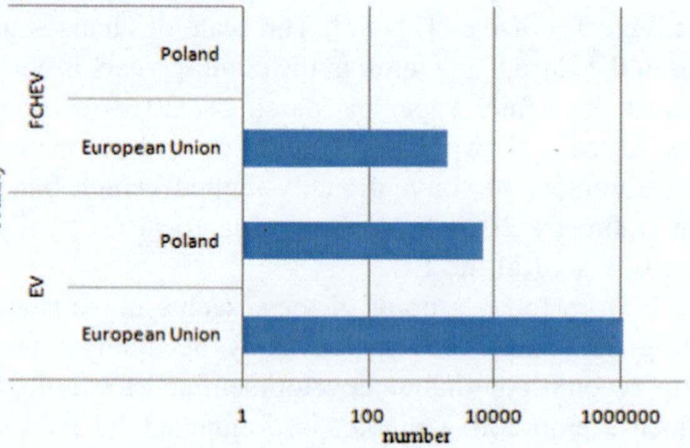

Fig. 4. Number of battery and hydrogen powered vehicles in the European Union and Poland in 2020 (Source: Own elaboration based on [20])

At present, 38 battery-powered models and 4 fuel-cell-powered models are offered on the Polish market [22]. Tables 1 and 2 show the models and the cost of purchasing electric vehicles in Poland.

Table 1. Models and cost of purchasing BEV electric vehicles in Poland (Source: Own elaboration based on [22–24]).

Model	Price [PLN]
Renaul Twizy	From 53 200
Skoda Citigo-e Iv	From 82 050
Smart Fortwo	From 96 900
Smart Forfour	From 98 400
Volkswagen e-up!	From 97990
Fiat 500	From 99500
Nissan Leaf	From 123900
Peugeot 208	From 124900
Renault Zoe	From 124900
Opel Corsa-e	From 132490
Citroen e-C4	From 134900
Mini Mini	From 139900
Opel Mokka-e	From 139900
Kia e-Soul	From 139900
Kia e-Niro	From 146990
Fiat 500C	From 140500
Mazda MX-30	From 142900
Peugeot 2008	From 149400
Honda e	From 154900
Volkswagen ID.3	From 155890
Hyundai Kona Electric	From 155900
DS 3 Crossback	From 155900
Volkswagen e-Golf	From 164890
Volkswagen ID.3 1 ST	From 167190
BMWi3	From 169700
Skoda Enyaq iV	From 182300
Hyundai IONIQ	From 185200
Audi Q4e-tron	From 195100
Mercedes EQA	From 199900
Tesla model 3	From 199900
Volkswagen ID.4	From 202390
BMW iX3	From 268900
Mercedes EQC	From 299000
Audi e-Tron	From 308400
Jaguar I-Pace	From 359500
Porsche Taycan	From 389000
Tesla S	From 395990
Tesla X	From 434990

When considering the total costs associated with purchasing and operating a vehicle, the lower costs will be for BEVs. For hydrogen-powered vehicles, not only is the purchase cost is higher, but also the cost of operating the vehicle, including the higher cost of refueling.

Table 2. Models and price of FCHEV vehicles in Poland (Source: Own elaboration based on [22, 25, 26]).

Model	Price [PLN]
Honda Clarity	From 265 000
Hyundai ix35	From 183 000
Hyundai Nexo	From 300 000
Toyota Mirai	From 299 900

2.1 Infrastucture

An important factor in the development of electromobility is the charging/fueling infrastructure for electric vehicles. Sufficiently developed infrastructure makes it possible to freely move such vehicles. In 2020, in the European Union there will be 19 9250 charging stations for low-power BEVs and 24 987 for high-power vehicles. The number of hydrogen charging stations in the EU is 125. In Poland, the number of charging stations for low-power BEV was 1039 units, while for high-power vehicles it was 652 units. The number of hydrogen vehicle charging stations in the EU was 0 units [21]. According to the data, in April 2021, there were 1456 BEV charging stations in Poland, of which 972 were free AC (alternating current) charging stations, while the

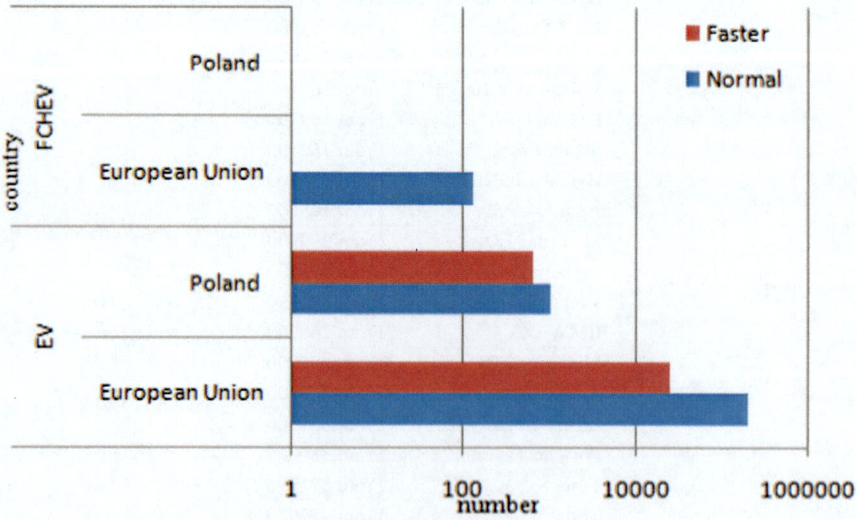

Fig. 5. Number of charging stations (low and high power) for EVs and hydrogen vehicles in EU and Poland in 2020 (Source: Own elaboration based on [20])

high-power DC (direct current) stations were 484. Figure 5 shows the number of charging stations (low and high power) for BEVs and hydrogen vehicles in Poland and the EU in 2020.

In Poland, the largest operators of electric vehicle charging stations are: GreenWay, PGE, Tauron, Orlen, Innogy, Rawicom, EV+, GO+Eauto, Ener-ga, Zepto, Ekoen, Ionity [27]. There is currently one hydrogen vehicle charging station operating in Poland. It is located in Warsaw and operated by a private company. The closest FCHEV filling stations are located in Berlin, Dresden (Germany) and Vienna (Austria). The first charging stations are planned to be built in Wroclaw, Warsaw and Gdansk [28, 29].

Currently, building a hydrogen vehicle refueling station and distribution network is very expensive. The cost of building a hydrogen station is estimated to be between 0.8 and 2.1 million Euros, while the pipeline infrastructure is between 0.4 and 3.2 million Euros.

Meanwhile, the costs of building charging stations for battery-powered vehicles are much lower. According to ORPA.PL Alternative Fuels Market Observatory, the cost of building a low-power charging station is 8846 EUR, a high-power station - from 17 693 to 26 536 EUR, the cost of electrical installation from 4422 to 15 478 EUR. In total, the construction of high-power stations is from 22 130 to 42 015 Euros.

2.2 Hydrogen Refuelling/Electricity Charging Cost

Due to the lack of filling stations for hydrogen vehicles in Poland, it is difficult to determine the cost of refuelling. In Germany, the cost of refuelling 1 kg of hydrogen is about 10 euros, i.e. 45.8 PLN per 100 km, while the cost of driving 100 km with an electric vehicle varies and depends on a number of factors, including the operator, offered tariffs, type of chargers (normal power, high power). At public stations the cost of 1 kWh ranges from 1.1 to 3.5 PLN depending on the type of charger. The cost of charging a battery-powered vehicle at an AC slow charging station ranges from 1.1 to 1.8 PLN per 1 kWh, while at a DC high-power station it ranges from 2 to 3.5 PLN.

Tables 3 and 4 below present the cost of driving a distance of 100 km for various electric vehicles and charging service providers currently in force in Poland.

Table 3 also presents the cost of charging a BEV from an electric socket available e.g. in a garage (tariff G11 and G12).

Table 3. Expected cost of charging electricity for 100 km traveled in Poland [PLN].

Brand	Model	Charging service provider				
		G11	G12	Ionity	GreenWay	
					AC	DC
Hyundai	Ioniq electric	8.09	3.67	51.5	16.8	21.9
Volkswagen	E-golf	9.52	4.32	60.6	19.7	25.8
BMW	I3	9.57	4.35	60.9	19.8	25.9
Renault	Zoe	11.17	5.07	71.1	23.1	30.3
Nissan	Leaf	11.28	5.12	71.8	23.4	30.5
Tesla	S P90D	13.20	6.00	84.0	27.4	35.7
Tesla	100D	13.20	6.00	84.0	27.4	21.9

Table 4. The cost of charging electricity per 100 km traveled in Poland [PLN].

Brand	Model	Charging service provider							
		Tauron		EV+		GO +Eauto		Elocity	Zepto
		AC	DC	AC	DC	AC	DC	AC	AC
Hyundai	Ioniq electric	23.5	32.5	17.6	26.5	16.9	23.5	14.7	14.7
Volkswagen	E-golf	27.7	28.2	20.8	31.1	19.9	27.7	17.3	17.3
BMW	I3	27.8	38.5	20.9	31.3	20.0	27.8	17.4	17.4
Renault	Zoe	32.5	44.9	24.4	36.5	23.3	32.5	20.3	20.3
Nissan	Leaf	32.8	45.3	24.6	36.9	23.6	32.8	20.5	20.5
Tesla	SP90D	38.4	53.0	28.8	43.2	27.6	38.4	24.0	24.0
Tesla	100D	38.4	53.0	28.8	43.2	27.6	38.4	24.0	24.0

2.3 Fueling/Charging Time for Electric Vehicles

The speed of charging a vehicles is an issue in which hydrogen vehicles are leading the way. In their case, refueling takes only a few minutes. For this reason, such vehicles can be a very good alternative to BEVs for longer distances. In the case of pure battery electric vehicles, the time it takes to recharge depends primarily on what type of charger is used to charge the electric vehicles and the charger built into the vehicle. It can take as long as 20 h to recharge a BEV if charging is done at home from a traditional electrical outlet. When using a fast charger or wallbox dedicated to a specific model, this time is reduced to a few hours. The fastest way to charge an electric vehicles will be to use a fast charging station available at a public place – in this way you can charge a battery in about 30 min. Table 5 shows the charging time of the BEV depending on the charging method.

Table 5. Nissan Leaf 40 kWh charging time (Source: Own elaboration based on [30, 31]).

Charging type	Time [h]	Charge level [%]
Home charging (230 V)	21	
Charging power 3.6 kW	13	From 0 to 100
Charging power 6.6 kW	7.5	From 0 to 100
Charging power 22 kW	1.5	From 20 to 80
Charging power 50 kW	0.4 to 0.6	From 20 to 80

In contrast, the refueling time for a 3.3 kg hydrogen Toyota Mirai is 3 min 55 s [32].

3 BEV and FCHEV Electric Vehicles

BEVs are electric vehicles that have an electric motor that is powered by batteries that store electricity. The batteries are charged by converting AC electricity into DC. These types of vehicles can travel between 100 and 540 km on a single charge, depending on battery capacity.

Meanwhile, FCHEVs are vehicles that have fuel cells instead of a battery that stores electricity. The fuel cell used is hydrogen. The range of FCHEVs is much longer than BEVs and is up to 500 km, hence they are preferred for longer routes [32]. Figure 6 shows the block diagram of the battery electric vehicle and Fig. 7 shows the FCHEV fuel cell vehicle.

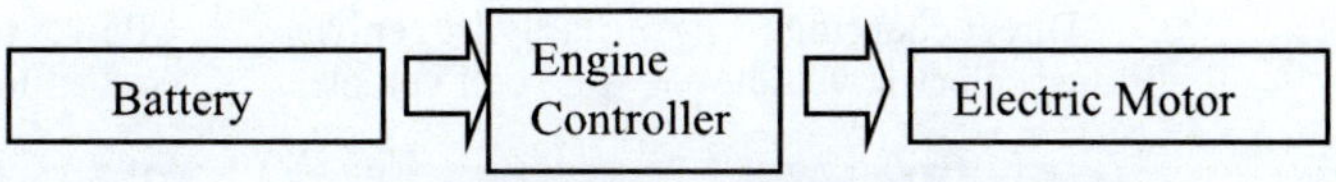

Fig. 6. Block diagram of a battery-powered electric vehicle (Source: Own elaboration based on [33])

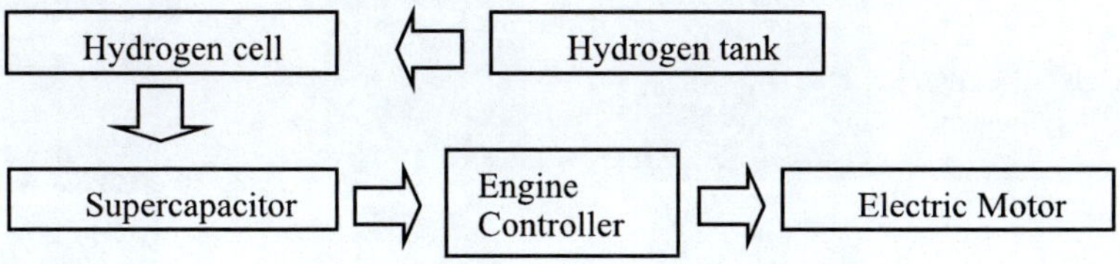

Fig. 7. Block diagram of a fuel cell vehicle (Source: Own elaboration based on [33])

One difference between these types of vehicles is energy storage. As the name implies, battery-powered vehicles use the stored energy in batteries for power. Hydrogen vehicles, on the other hand, are equipped with a high-pressure tank that stores compressed hydrogen. A tank containing hydrogen at several times higher pressure (600 bar) must be used. This solution brings problems related to the energy needed to compress hydrogen at the tank manufacturer and the fact that the hydrogen particles are small, in combination with air creates an explosive mixture, storage must be in a sealed and made of a suitable coating. Compression is a process that consumes a lot of energy. It increases the energy used by half than transporting hydrogen in this way. This contributes to an energy drop and increases the cost of producing the tank. The hydrogen travels to the fuel cells, reacting with oxygen to produce electricity and steam. Analyzing both production solutions, fuel cell production is very expensive. This is related to the construction materials from which it is made and the larger number of components that go into the vehicle. The construction of a hydrogen vehicle also requires additional power supply in the form of a battery [34–37].

3.1 Energy Efficiency

The non-governmental organisation Transport & Environment has determined that of the self-driving vehicles with different propulsion systems, the electric BEV is the most energy efficient. To simplify the calculations, it assumed that in all external processes, the additional electricity did not come from fossil fuels (coal) or nuclear power plants, but from completely renewable sources (wind, sun). On the basis of the carried out analyses, it has been shown that an electric vehicles charged with photovoltaic panels is capable of using effectively 73% of the energy supplied, while a vehicles with hydrogen cells uses only 30% (Fig. 8).

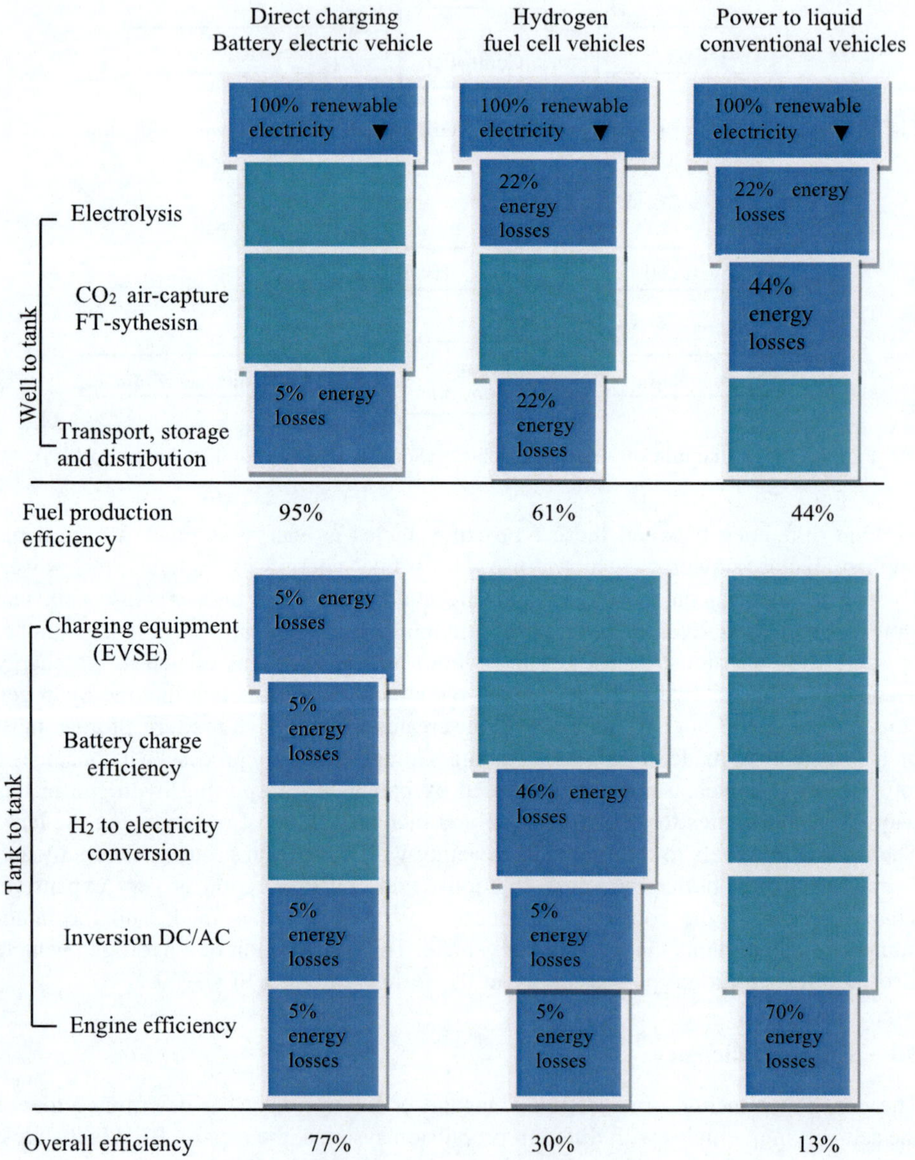

Fig. 8. Efficiency of different passenger vehicles technology pathways based on renewable electricity (Source: elaboration based on [38])

Already at the stage of hydrogen extraction by electrolysis, as much as 45% of the initial energy is lost. Subsequent stages of the process include compression, condensation, transport, filling and energy generation in the fuel cell. By the time the hydrogen is converted to electricity in the vehicle, just over half of the remaining energy is lost.

4 Environmental Impact of Electric Vehicles

An important aspect of battery electric vehicles and fuel cell electric vehicles is their impact on the environment. The main advantage of electric vehicles is the non-pollution operation. Hydrogen vehicles produce electricity, water and heat and therefore do not have a negative impact on the environment. Due to the lack of emission of toxic exhaust components into the atmosphere, global warming is reduced. Battery-powered vehicles, on the other hand, are zero-emission during operation. This is due to the fact that BEVs use electric energy stored in batteries [39].

The problem may be the methods of obtaining the power source for electric vehicles. Using hydrogen fuels and the hydrogen used for power is environmentally friendly, but the processes for obtaining hydrogen can be harmful to the environment and energy intensive [40].

Currently, the most common methods for obtaining hydrogen are [41]:

- electrolysis of water,
- natural gas reforming,
- coal or coke gasification,
- electrolysis water.

Figure 9 shows the percentage of hydrogen extraction methods.

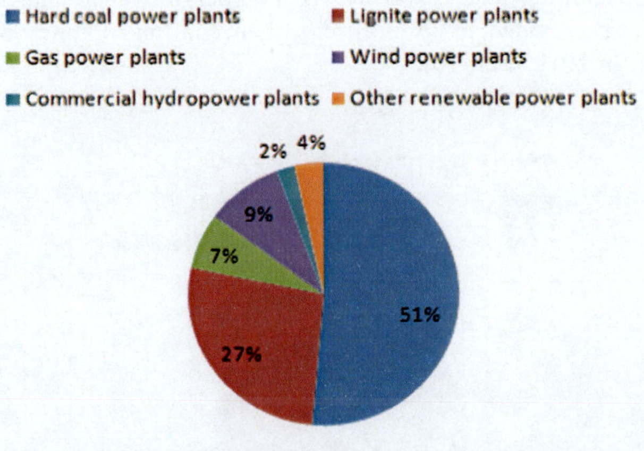

Fig. 9. Percentage of methods in hydrogen extraction (Source: Own elaboration based on [41])

When analyzing methods for obtaining hydrogen, it should be noted that greenhouse gases are emitted during production. Steam reforming, autothermal and gasification methods require high temperatures. To obtain them, fossil raw materials are burned, which has the side effect of emitting carbon dioxide. Carbon dioxide emissions also occur during the reactions necessary to extract hydrogen. Another pollution risk is the release of carbon dioxide during the extraction of fossil fuels from underground deposits.

Therefore, the use of fossil fuels to produce hydrogen is expensive and can be detrimental to the environment. Producing hydrogen from water is also not good for the environment. Water electrolysis uses a lot of water to produce hydrogen. Also the production of hydrogen from biomass has its limitations due to the surface area. Therefore, the production of hydrogen may have an adverse effect on the environment, contributing to the fact that the hydrogen vehicle will not be an ecological solution [36, 42].

In Poland, hydrogen extraction is done using coking and steam reforming. These are energy-intensive methods using fossil fuels, thus contributing to carbon dioxide emission [43, 44]. Electric vehicles, due to using only an electric motor while driving, convert electricity stored in batteries. The conversion of electrical energy into mechanical energy involves only the emission of small amounts of heat resulting from losses during energy conversion, so the operation of the electric vehicles drive is characterized by very high environmental cleanliness.

Their overall CO2 (carbon dioxide) and pollutant emissions are limited to those of the electricity source. Thus, BEVs do not emit emissions at the point of use, which leads to a shift from "tailpipe" emissions to those generated at power plants. In this situation, the total emissions of BEVs are typically far removed from cities, with only a minor impact on the health and comfort of most residents.

Studies show that electricity derived only from coal means that BEV electric self-driving vehicles will contribute more greenhouse gas emissions than internal combustion engine vehicles [45–47].

Meanwhile, according to a report by Agora Energie-wende and Sandbag, wind, solar, and biomass provided more electricity than coal and lignite combined in the European Union in 2019 (Fig. 10).

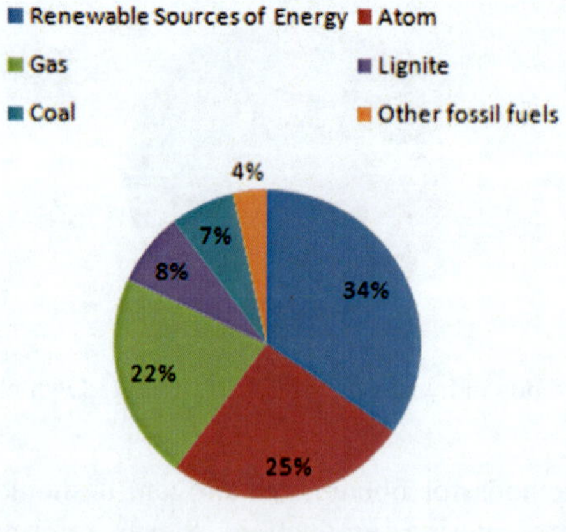

Fig. 10. EU electricity generation in 2019 (Source: Own elaboration based on [48])

This means that vehicles powered solely by electricity from the average European basket will reduce GHG (greenhouse gas) emissions, althought slightly, compared with vehicles powered by both diesel and gasoline engines, assuming a vehicle life of 150 000 km [45–47]. As it turns out [49], an electric Nissan Leaf (80 kW, 280 Nm) emits slightly less CO2 per kilometer than an internal combustion Toyota Auris (diesel 1.4, 66 kW, 205 Nm). Conversion of the vehicle's energy consumption into emissions by Polish power plants shows emissions of 118 g/km, while for the Toyota Auris it is 124 g/km. This data shows a small but nevertheless advantage of an electric vehicle.

According to a Swiss study, a combustion vehicle would have to burn three to four liters of fuel to be as environmentally friendly as an electric vehicle (with the same characteristics) powered by a power plant emitting the average European amount of pollution. The data is even more favorable for an electric vehicle if it uses renewable energy or zero-emission sources.

5 Conclusions

Currently, the battery-powered vehicle market in Poland is performing much better than the cell-powered vehicle market. Mass production of hydrogen vehicles is currently at a very early stage of development as evidenced by the very small share of these vehicles in the motorization. The small number of hydrogen vehicle models offered may cause the development of the hydrogen vehicle market to be insignificant. Another barrier to hydrogen vehicles is their price. The hydrogen vehicles offered today are much more expensive than the most commonly purchased battery-powered vehicles. When analyzing the performance of a battery-powered vehicle and a cell-powered vehicle, the latter performs worse. Battery-powered vehicles have efficiencies of up to 80%, while the efficiency of hydrogen vehicles is only 45%. Another very significant barrier to the growth of the cell powered vehicle market is the infrastructure and processes for obtaining hydrogen. Analyzing the infrastructure of battery powered vehicles in Poland, it is sufficiently developed. However, the refuelling infrastructure of hydrogen vehicles is at a completely different stage. At the moment there are no stations of this type available in Poland. A problem in the development of hydrogen vehicle charging infrastructure is the high cost of building the stations and building a hydrogen transmission network. Also, storage and refueling of hydrogen vehicles is problematic. This is due to the high compression of hydrogen, which requires the use of electricity. The second and very big disadvantage is the possibility of hydrogen leaking into the atmosphere and consequently destroying the ozone layer, with which hydrogen has a chemical reaction.

The transportation of hydrogen is also complicated, in special tanks which are covered with special coatings that prevent hydrogen from penetrating the atmosphere. Transporting hydrogen via pipelines is the most beneficial method. However, the costs of building the entire network are high.

The biggest disadvantage of hydrogen vehicles over battery-powered vehicles is the extraction of hydrogen. Using steam methane reforming has a negative impact on the environment. Firstly, it uses fossil fuels, thus depleting resources. Secondly, this method requires large amounts of energy. As a result, the production of 1 kg of hydrogen during steam reforming emits 5.5 kg of carbon dioxide into the atmosphere. The use of coal as a feedstock is even more detrimental to the environment, as the production of hydrogen emits twice as much carbon dioxide into the atmosphere. The use of hydrolysis also has a negative impact because this process is very energy intensive. Extracting hydrogen by hydrolysis consumes twice as much energy as the hydrogen extracted by this method contains. Using renewable energy sources to extract hydrogen is also not the best solution. Because a better solution is to transfer the energy immediately to battery-powered vehicles [34]. Current conditions indicate that currently hydrogen vehicles are not a better solution than battery powered vehicles in shaping a sustainable transport system. The production of hydrogen and its negative impact on the atmosphere and the environment. The use of renewable energy sources as one of the main producers of electricity gives a better perspective to the use of battery electric vehicles.

References

1. Jodkowska, L.: Individual settings and distinctions in awards. Trends and Distinctive Awards. Zapol, Szczecin (2001)
2. Mazur-Wierzbicka, E.: The place of sustainable development in Polish and EU ecological policy at the beginning of the 21st century. Economics **8**(1), 317–328 (2006)
3. Nikołajewka, A., Adey, P., Cresswell, T., Lee, J.Y., Nóvoa, A., Temenos, C.: Commoning mobility: towards a new politics of mobility transitions. Trans. Inst. Br. Geogr. **44**, 346–360 (2019). https://doi.org/10.1111/tran.12287
4. Correa, D.F., et al.: Towards the implementation of sustainable biofuel production systems. Renew. Sustain. Energy Rev. **107**, 250–263 (2019)
5. Standing, C., Standing, S., Biermann, S.: The implications of the sharing economy for transport. Transp. Rev. **39**(2), 226–242 (2019). https://doi.org/10.1080/01441647.2018.1450307
6. Pisonia, E., Christidis, P., Thunis, P., Trombetti, M.: Evaluating the impact of "Sustainable Urban Mobility Plans" on urban background air quality. J. Environ. Manage. **231**, 249–255 (2019)
7. European Enviroment Agency, https://www.eea.europa.eu
8. European Union: European low-emission mobility strategy, COM (2016) 501 (2016). https://www.europarl.europa.eu/doceo/document/TA-8-2017-0503_PL.html. https://doi.org/10.1016/j.jenvman.2018.10.039
9. Eurostat. Statistics Explained: Greenhouse gas emissions by IPCC source sector, EU-28, png. (2017). https://ec.europa.eu/eurostat/statistics-explained/index.php?title=File:Greenhouse_gas_emissions_by_IPCC_source_sector,_EU-28,_2017.png&oldid=461257
10. European Automobile Manufacturers: Association, https://www.acea.be/
11. Ahn, S.W.: EU member-states' reaction to the climate change in transportation: French case of the development of electric vehicles. J. Contemp. Eur. Stud. 103–127 (2019)
12. Astegiano, P., Fermi, F., Martino, A.: Investigating the impact of e-bikes on modal share and greenhouse emissions: a system dynamic approach. Transp. Res. Procedia **37**, 163–170 (2019)
13. Lu, M.J., Taiebat, M., Xu, M., Hsu, S.-C.: Multiagent spatial simulation of autonomous taxis for urban commute: travel economics and environmental impacts. J. Urban Plann. Dev. **144**(4), 04018033 (2018). https://doi.org/10.1061/(ASCE)UP.1943-5444.0000469

14. White Paper on transport, Roadmap to a single European transport area – towards a competitive and resource-efficient transport system. https://ec.europa.eu/transport/sites/transport/files/themes/strategies/doc/2011_white_paper/white-paper-illustrated-brochure_en.pdf
15. Breakthrough of electric vehicle threatens European car industry – ING Economics Depart ment. https://www.ing.nl/media/ING_EBZ_breakthrough-of-electric-vehicle-threatens-European-car-industry_tcm162-128687.pdf
16. Buczaj, M.: The use of alternative power sources for vehicles in the light of EU standards and directives on the example of Poland.Motrol. In: Motorization and Power Industry in Agriculture, vol. 6, pp. 12–19 (2006)
17. Murawski, J., Szczepański, E.: Prospects for the development of electromobility in Poland. Logistics **4**, 2249–2258 (2014)
18. Śliwka, M., Łyko, P., Pomykała, R.: Economic and ecological aspects of selected alternative sources of power supply for passenger cars, Logistics (2015). http://www.zarz.agh.edu.pl/adubel/wp-content/uploads/2015/09/Logistyka-4-2015.pdf
19. Polakowski K.: Electric cars are the vehicles of the near future? Works of the Electrotechnical Institute, vol. 252, pp. 19–39 (2011)
20. European Alternative Fuels Observatory, www.eafo.eu
21. For the first time in Poland, 6 hydrogen cars have been registered, https://globenergia.pl/po-raz-pierwszy-w-polsce-zarejestrowano-6-samochodow-wodorowych/
22. Electric car prices, https://www.motonews.pl/auta-nowe/elektryczne/
23. Electric cars available in Poland, https://orpa.pl/elektromobilni-pl-internetowy-katalog-pojazdow-elektrycznych-dostepnych-w-polsce/
24. How much does an electric car cost? We have collected all the models available in Poland – market overview, https://moto.pl/MotoPL/7,176355,26723314,ile-kosztuje-samochod-elektryczny-zebralismy-wszystkie-modele.html
25. I drove a hydrogen Hyundai Nexo in Poland, which cannot be refueled here, https://e.autokult.pl/33445,hyundai-nexo-tankowanie-cena-pierwsza-jazda-w-polsce
26. Hyundai ix35 Fuel Cell – the first production hydrogen car, https://www.motofakty.pl/artykul/hyundai-ix35-fuel-cell-pierwszy-seryjny-samochod-na-wodor-film.html
27. Ranking of the largest networks of electric car charging stations in Poland, https://wysokienapiecie.pl/26820-ranking-najwiekszych-sieci-stacji-ladowania-samochodow-elektrycznych-w-polsce/
28. The first hydrogen refueling station in Poland has been built in Warsaw, there are also Toyota Mirai and Hyundai Nexo, https://elektrowoz.pl/ladowarki/pierwsza-w-polsce-stacja-tankowania-wodoru-stanela-w-warszawie-sa-tez-toyoty-mirai-i-hyundaie-nexo-tylko-u-nas/
29. The first hydrogen refueling station in Poland is getting closer. Gdańsk, not Warsaw, https://elektrowoz.pl/ladowarki/pierwsza-stacja-tankowania-wodoru-w-polsce-coraz-blizej-gdansk-nie-warszawa/
30. How long does it take to charge an electric car? We check the available methods, https://moto.pl/MotoPL/7,176355,26268398,ile-trwa-ladowanie-samochodu-elektrycznego.html
31. How long does it take to charge an electric car?, https://electronite.eu/ile-trwa-ladowanie-samochodu-elektrycznego/
32. Electric cars and hydrogen as the "fuel of the future". How is it with the speed of refueling and loading, https://elektrowoz.pl/porady/samochody-elektryczne-a-wodor-jako-paliwo-przyszlosci-jak-to-jest-z-ta-szybkoscia-tankowania-i-ladowania-liczymy/
33. Tyczka, M., Skarka, W.: Electric car powered by a hydrogen cell. The alternative to battery power. In: XIII ProCAx Engineering Forum 2015, vol. 3, pp. 238–239. Mechanic (2016)
34. Hydrogen cars as the future of the automotive industry?, http://www.ekspert-flotowy.pl/artykuly/594/auta-wodorowe-przyszloscia-motoryzacji/

35. Fuel Cells, http://zs9elektronik.pl/inne/karolina/podr_5_ogniwa_paliwowe.pdf
36. Marzec, A.: Problems of hydrogen fuel. Energy Policy **1**(1), 89–96 (2007)
37. Folentarska, A., Kulawiak, D., Ciesielski, W., Pavlyuk, V.: Modern materials for storing hydrogen as the fuel of the future. Chem. Environ. Biotechnol. **19**, 125–130 (2016)
38. Transport & Environment, https://www.transportenvironment.org/
39. Kamiński, T., Filipek, P.: Hydrogen fuel cells – ecological source of DC power supply to vehicles. Adv. Sci. Technol. **8**, 20–28 (2011)
40. Auto Center, https://www.autocentrum.pl/newsy/czy-na-pewno-ogniwa-paliwowe-sa-przyjazne-srodowisku
41. Wiącek, D.: Hydrogen as the Fuel of the Future, pp. 446–451. Buses (2011)
42. Uhl, T.: Is hydrogen the future of urban transport? Hydrogen refueling stations. Part 1. New Energy, pp. 81–87 (2020)
43. Ściążko, M., Smółka, B., Wnecki, T.: Hydrogen perspectives in transport. New Energy (2018)
44. How European Hydrogen Generators Implement Zero-Emission Production Technologies, https://300gospodarka.pl/explainer/jak-wytworcy-wodoru-w-europie-wdrazaja-zeroemisyjne-technologie-produkcji
45. Ellingsen, L.A.W., Singh, B., Strømman, H.: The size and range effect: lifecycle greenhouse gas emissions of electric vehicles. Environ. Res. Lett. **11**, 1–8 (2016)
46. Miotti, M., Hofer, J., Bauer, C.: Integrated environmental and economic assessment of current and future fuel cell vehicles. Int. J. Life Cycle Assess. **22**(1), 94–110 (2015)
47. Notter, D.A., Kouravelou, K., Karachalios, T., Daletou, M.K., Haberland, N.T.: Life cycle assessment of PEM FC applications: electric mobility and μ-CHP. Energy Environ. Sci. **8**, 1969–1985 (2015)
48. High voltage, https://wysokienapiecie.pl/26386-produkcja-energii-elektrycznej-z-wegla-w-europie-ostro-hamuje/
49. Solaris, http://solaris18.blogspot.com/2009/12/czy-samochody-ektryczne-sa.html

Application and Comparison of Machine Learning Algorithms in Traffic Signals Prediction

Anastasiia Kunashko[1], Feng Xie[1](✉), Sebastian Naumann[1], Yin Gao[2], and Jun Li[2]

[1] Institute of Automation and Communication, Magdeburg, Germany
{feng.xie, sebastian.naumann}@ifak.eu

[2] Quanzhou Institute of Equipment Manufacturing, CAS, Quanzhou, China

Abstract. Traffic signals prediction based on machine learning plays a vital role in future vehicle to infrastructure communication in intelligent transport systems, since they can help vehicles avoid waiting and reduce energy consumption with a higher communication security compared with traditional radio-based pattern. In this work, Baseline Model, Linear Model, Dense Model, Convolutional Neural Network, Long Short Term Memory (LSTM) and Autoregressive LSTM are applicated to forecast the traffic signals based on the historical time series. After comparison, LSTM outperforms other methods with a high potential to be extended further for a better accuracy.

Keywords: Traffic signals · Time series forecasting · Machine learning · V2I

1 Introduction

Due to an explosion of communication requirements, it seems that public transportation can not be guaranteed always to obtain priority to drive through intersections via analog radio-based communication between traffic lights, especially when there are competing requests. In intelligent transport systems (ITS), this could be solved by vehicle to infrastructure (V2I) communication based on standards of ETSI ITS-G5 and ISO/TS 19091. However, it will cause data security problem and high cost of complex systems implemented with large amounts of communication modules in ITS. Therefore, in order to avoid such a heavy dependency on the direct communication with infrastructures, AI-based prediction of the future signals based on historical data should be a suitable choice.

The prediction of upcoming traffic signals could be viewed as a kind of time series forecasting problem. Varieties of machine learning models have been applicated in prediction problems. Elsworth and Güttel [1] proposed a combination model of a recurrent neural network (i.e. Long Short Term Memory (LSTM) Network) with a dimension-reducing symbolic representation to overcome limitations in time series forecasting, such as high sensitivity to the random original weights and to the hyper parameters. In order to solve the uncertainties caused by state variables, which can be only observed as irregular epochs in time, Bhattacharjya et al. [2] developed a novel event-driven continuous time Bayesian network to model the cases when state variables

E. Macioszek and G. Sierpiński (Eds.): TSTP 2021, LNITI, pp. 69–78, 2022.
https://doi.org/10.1007/978-3-030-93370-8_5

are influenced by the occurrence of different events. Their approach adopted a greedy search procedure for structure learning, which was proved to be also applicable for limited data set. Two of the most commonly used deep learning methods for time series forecasting are recurrent ad convolutional neural networks [3]. However, Dabrowski et al. [3] put forward that these time-invariant architecture will result in limited performance in multi-step-ahead forecasting. Therefore, they proposed ForecastNet, which is a kind of time-variant deep feed-forward neural network architecture. Aiming to deal with forecasting problems for multiple short time series, Shi et al. [4] developed a novel approach based on the method of autoregressive integrated moving average. They applied Tucker decomposition to map higher-order block Hankel tensors to compressed core tensors. As a result, the forecasting accuracy was improved and calculating cost was successfully reduced. To deal with irregular non-stationary signals, Zhang and Zhou [5] raised the Harmonic Recurrent Process, which integrates recurrent period-varying patterns in harmonic analysis. Oreshkin et al. [6] built a meta-learning framework to forecast zero-shot time series, which can train a neural network on a source time series while can be applicated to another target time series without retraining. Since the model selection for positive unlabeled time series classification is a new research area, Liang et al. [7] designed a model selection framework with active learning. Based on the model selection framework, a model performance evaluation strategy and several active learning sampling strategies were proposed. For multivariate time series, where variables influence each other, existing methods have limited performance to deal with the internal dependencies of variables. Therefore, Wu et al. [8] designed a general Graph Neural Network to handle the relational dependencies. Obviously, time series forecasting is a really wide topic for research. Though the traffic signals are also a kind of time series, it is still difficult to select a suitable model for traffic area based on V2I communication. In order to provide a general forecasting method of traffic signal, this paper will try to applicate and compare different basic models and then make a suggestion of model selection for the first step.

The rest of this paper is organized as follows. Section 2 introduces how the data of traffic signals should be prepared, including the generation and further processing. Section 3 explains the models applicated in this work. The compared results are described in Sect. 4. Section 5 concludes the work of model selection for future signals forecasting.

2 Preparation of Data

2.1 Data Generation

For training machine learning algorithms, a data set of traffic signals is necessary. In this work, a three-legs intersection is simulated for the generation of data. As shown in Fig. 1, Dn1 ~ Dn represent detectors at specific lanes of the intersection, if the values are true, it means there are cars in queue. Labels of K1 ~ Kn represent the status of the traffic signals (i.e. RED, GREEN, YELLOW, YELLOW_RED). The conflicts between different lanes are also described in Fig. 1, which means the switching time of signals should avoid these conflicts between K1 & K3, K1 & K4, K2 & K3, K2 & K4, and K3 & K4. The simulation is proceeded based on the following assumptions:

- in order to avoid conflicts, there should be a safety time between a signal changing to RED and the conflict signal switching to GREEN. The safety time in this work is set to be 3 s,
- for GREEN signals, the minimal duration is set to be 10 s, while the maximal duration is 40 s,
- as long as the detector indicates ON, the GREEN signal will be extended, at most reaching the maximal duration.

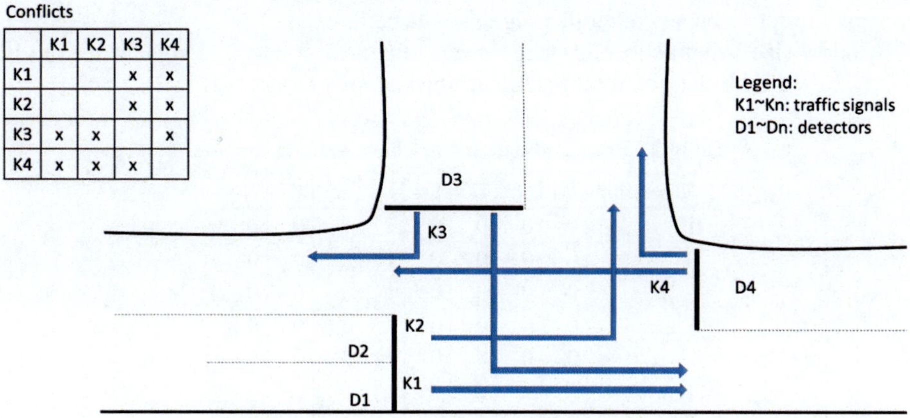

Fig. 1. Three-legs intersection for data collection

2.2 Data Processing

After simulation, the generated data is presented in Table 1 with 10001 data in total, where D1 ~ D4 are attribute values and K1 ~ K4 are target values. In order to train the data for machine learning methods, these data have to be further processed.

Table 1. Original generated data of traffic signals.

Timestamp	D1	D2	D3	D4	K1	K2	K3	K4
0	false	false	false	false	GREEN	GREEN	RED	RED
1	false	false	false	false	GREEN	GREEN	RED	RED
2	false	false	false	false	GREEN	GREEN	RED	RED
3	false	false	false	false	GREEN	GREEN	RED	RED
4	false	false	false	false	GREEN	GREEN	RED	RED
…	…	…	…	…	…	…	…	…
9996	true	false	false	true	RED	RED	RED	RED
9997	true	false	true	true	RED	RED	RED	RED
9998	true	false	true	true	RED	RED	RED	RED
9999	true	false	true	true	YELLOW_RED	RED	RED	RED
10000	true	false	true	true	GREEN	RED	RED	RED

For the training set, 70% of the total data is selected for training, and 20% is selected to be the validation set, the rest 10% is selected as the test set. In this situation, m historical timestamps with signal status will be given, n steps of signals need to be predicted ahead. Therefore, in this work, this problem is categorized as a problem of multi-step time series forecasting. All generated data have to be transformed to Table 2 based on the following assumptions:

- for every target attribute K1 ∼ K4, a separate model will be trained,
- only GREEN means a car can drive freely through the intersection, so GREEN signal is set to be 1 while other signals will be 0,
- window size is defined as 60, which means the next 60 steps of signals need to be forecasted with the given 60 historical timestamps of signals (Fig.2).

Table 2. Processed data for machine learning training.

Timestamp	D1	D2	D3	D4	K1	K2	K3	K4
0	0	0	0	0	1	1	0	0
1	0	0	0	0	1	1	0	0
2	0	0	0	0	1	1	0	0
3	0	0	0	0	1	1	0	0
4	0	0	0	0	1	1	0	0
…	…	…	…	…	…	…	…	…
9996	1	0	0	1	0	0	0	0
9997	1	0	1	1	0	0	0	0
9998	1	0	1	1	0	0	0	0
9999	1	0	1	1	0	0	0	0
10000	1	0	1	1	1	0	0	0

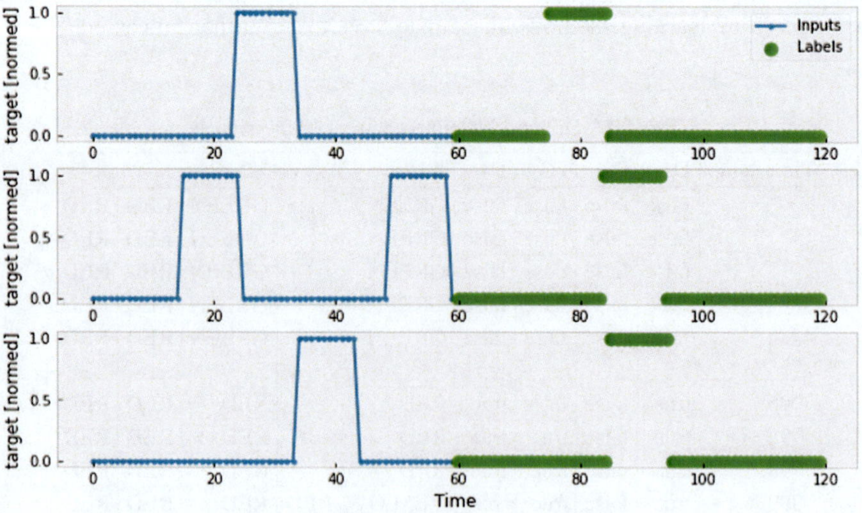

Fig. 2. Window set for data processing

3 Machine Learning Methods for Traffic Signal Forecasting

3.1 Benchmark Models

For a better comparison, it is necessary to applicate benchmark models. In this research, Baseline Model (BM) and Linear Model (LM) are adopted as benchmarks.

BM. This is a trivial model that ignores feature values and learns to predict the last seen values for each label. As shown in Fig. 3, BM can only capture nearly the changing tendency of the inputs.

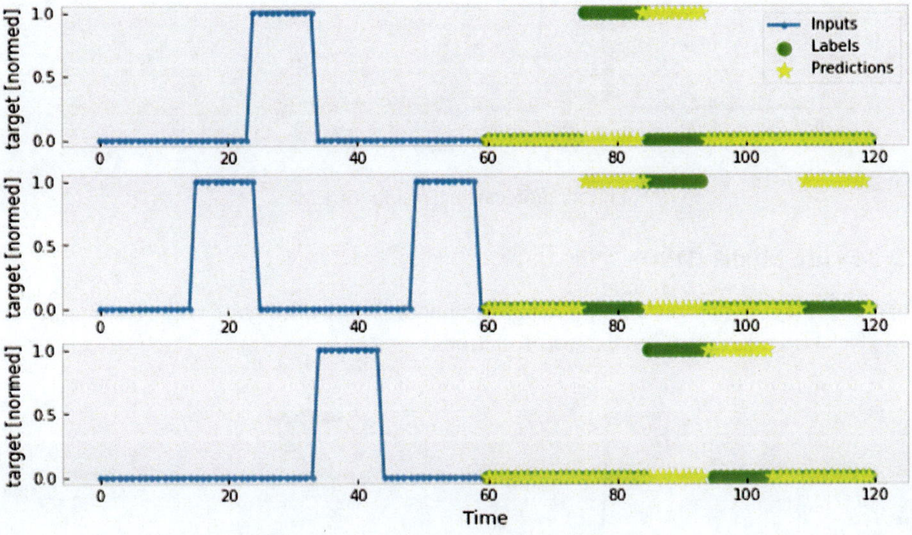

Fig. 3. Forecasting results of BM

LM. This is a model consisting of only one simple linear layer. As shown in Fig. 4, LM has a poor performance to capture the switching behavior of input signals, and always outputs values of 0.

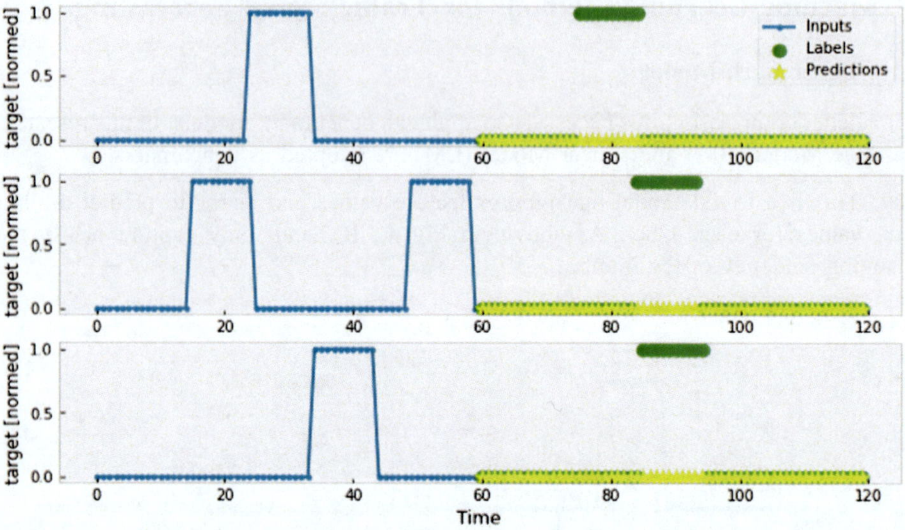

Fig. 4. Forecasting results of LM

3.2 Dense Model (DM)

This model includes three densely-connected neural network layers with rectified linear unit (ReLU) and sigmoid activation functions.

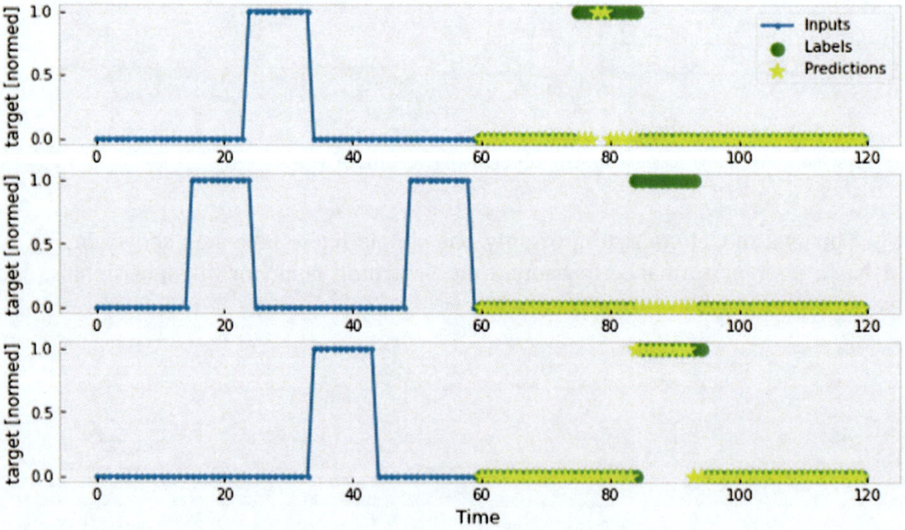

Fig. 5. Forecasting results of DM

As shown in Fig. 5, DM can forecast partially the future signals.

3.3 Convolutional Neural Network (CNN)

The CNN consists of one convolutional layer and two dense layers with ReLU and sigmoid activation functions. As depicted in Fig. 6, CNN performs a passable forecasting ability.

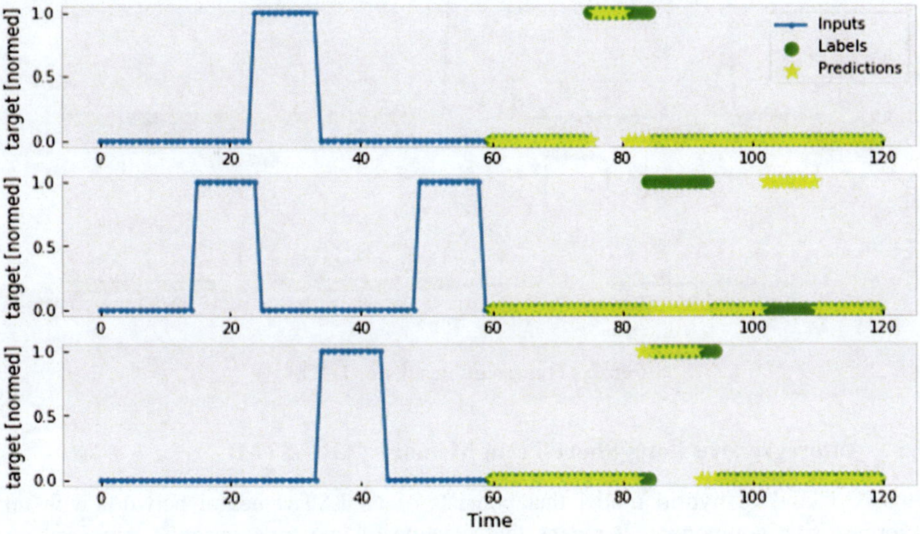

Fig. 6. Forecasting results of CNN

3.4 LSTM

This is a kind of recurrent neural network that consists of one LSTM layer and one dense layer with sigmoid activation function. The forecasting results are shown in Fig. 7, it seems LSTM can have a better performance when the future series have a similar shape of input series.

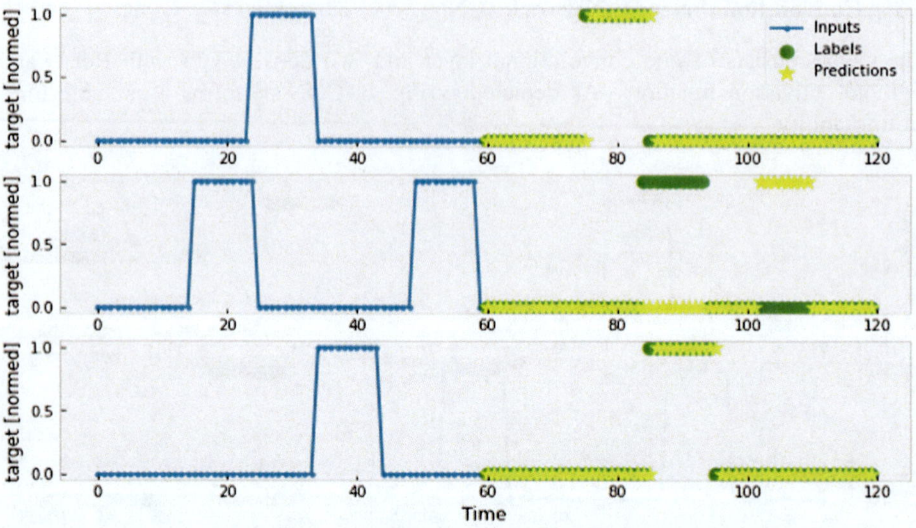

Fig. 7. Forecasting results of LSTM

3.5 Autoregressive Long Short Term Memory (AR-LSTM)

This is a kind of hybrid model that consists of an LSTM neural network with an autoregressive component. It means that a value of one time series is regressed on previous values in the same series, with the forecasting results shown in Fig. 8.

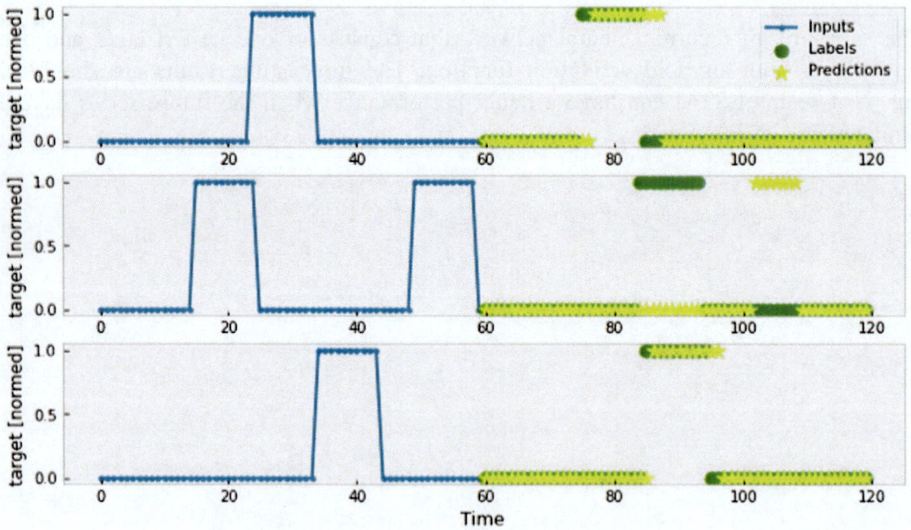

Fig. 8. Forecasting results of AR-LSTM

4 Comparison of Forecasting Results

The final forecasting results for K1 ~ K4 of mentioned machine learning methods are presented in Fig. 9. Both validation and testing sets have similar outputs, with the average binary accuracy calculated respectively. Obviously, all researched machine learning methods have stable and similar forecasting ability for signals at K1, K2, K3 and K4, which means this evaluation result for traffic signals is reliable. Though DM, CNN, LSTM, AR-LSTM have a higher accuracy than benchmark models, the AR-LSTM has a relatively unstable performance compared with other models.

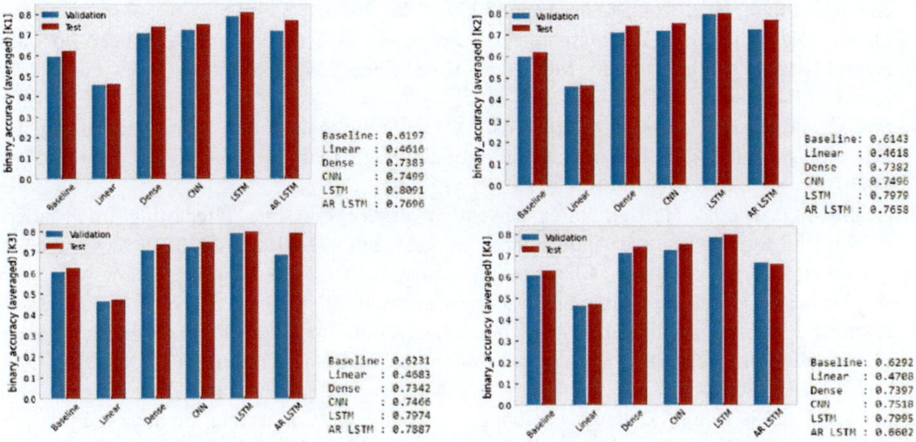

Fig. 9. Comparison of forecasting results of different machine learning models

5 Conclusions

For K1 to K4, LSTM has an accuracy of 0.8091, 0.7979, 0.7974 and 0.7995, which is approximately a forecasting accuracy of 80% based on several historical data. After a comparison, it is obvious that LSTM outperforms other machine learning models in this research. It is always necessary to compare the complex models with a simpler benchmark model to distinguish worse forecasting methods. In this research, LM has only an accuracy less than 50%, BM has an accuracy more than 60%. Though all other machine learning methods have better results, for K4, AR-LSTM performs a result very closed to BM, which proves that AR-LSTM has a kind of randomness with unstable performances. Therefore, a further extension of LSTM should be proceeded but it is better to avoid AR-LSTM in this case.

This is a very first step to explore suitable forecasting methods based on machine learning models for traffic signals. This work indicates a possibility to combine LSTM with other models to have a better forecasting result. And in the future work, e.g. for the real-time V2I communication in ITS, the received time series of traffic signals should at least include these features as a timestamp. For the next step, besides the historical data, internal commutative influences of neighbor traffic signals should also be researched for a more precise prediction.

References

1. Elsworth, S., Güttel, S.: Time Series Forecasting Using LSTM Networks: A Symbolic Approach (2020). https://arxiv.org/pdf/2003.05672.pdf
2. Bhattacharjya, D., Shanmugam, K., Gao, T., Mattei, N., Varshney, K., Subramanian, D.: Event-driven continuous time bayesian networks. In: Proceedings of the AAAI Conference on Artificial Intelligence, vol. 34, no 4, pp. 3259-3266 (2020)
3. Dabrowski, J.J., Zhang, Y., Rahman, A.: ForecastNet: A Time-Variant Deep Feed-Forward Neural Network Architecture for Multi-step-Ahead Time-Series Forecasting. https://arxiv.org/abs/2002.04155
4. Shi, Q., et al.: Block hankel tensor ARIMA for multiple short time series forecasting. In: Proceedings of the AAAI Conference on Artificial Intelligence, vol. 34, no 4, pp. 5758-5766 (2020)
5. Zhang, S.Q., Zhou, Z.H.: Harmonic recurrent process for time series forecasting. In: Frontiers in Artificial Intelligence and Applications, vol. 325, pp. 1714-1721 (2020)
6. Oreshkin, B.N., Carpov, D., Chapados, N., Bengio, Y.: Meta-Learning Framework with Applications to Zero-Shot Time-Series Forecasting (2020). https://arxiv.org/abs/2002.02887
7. Liang, S., Zhang, Y., Ma, J.: Active model selection for positive unlabeled time series classification. In: 36th IEEE International Conference on Data Engineering (ICDE), vol. 2020, pp. 361-372 (2020)
8. Wu, Z., Pan, S., Long, G., Jiang, J., Chang, X., Zhang, C.: Connecting the dots: multivariate time series forecasting with graph neural networks. In: Proceedings of the 26th ACM SIGKDD International Conference on Knowledge Discovery & Data Mining, pp. 753-763 (2020) https://doi.org/10.1145/3394486.3403118

Check for updates

Comprehensive Estimate of Life Cycle Costs of New Technical Systems, Using Reliability Verification Algorithm

Natalia Tereshina, Victor Podsorin, Nataliya Tereshina(✉), Pavel Metelkin, and Larisa Chernyshova

Russian University of Transport, Moscow, Russia
np.tereshina@ya.ru

Abstract. This article discusses major aspects of innovative activities making their contribution to increased competitiveness of transport companies through the implementation of innovation advantages and benefits in comparison with existing technical systems. The embodiment of nonlinear innovative process organization models expedites the implementation of scientific achievements in practice in the form of new technical systems characterized by higher reliability and availability indicators. Substantiation of decisions on implementing new technical systems based on the life cycle theory takes into account reliability indicators and defines the liability of the innovative process participants, first of all, developers and manufacturers of technical systems, for a failure to comply with target life cycle cost parameters. A comprehensive estimate of life cycle costs of new technical systems by using reliability indicators can form the basis of a professional liability insurance system for providing reliability indicator levels by developers, manufacturers and suppliers.

Keywords: Life cycle · Verification · Technical systems · Reliability · Transport

1 Introduction

Competitive success of manufacturing firms is by and large determined by the success of the products they introduce to the market. This is why companies continuously try to improve the efficacy of their product realization process. Product Lifecycle Management (PLM) is a business solution which aims to streamline the flow of information about the product and related processes throughout the product's lifecycle such that the right information in the right context at the right time can be made available. Yet, few organizations are positioned to reap the true benefits of PLM [1].

In recent years, the Russian business community also increasingly uses a life cycle theory based approach [2–4]. Similar trends are observed in the Russian transport segment [5]. While upgrading infrastructure facilities, rolling stock and process equipment, transport companies adhere to a life cycle concept in business management [6].

Nowadays, the global scientific community is widely discussing the ideas of applying the life cycle theory in practice and is studying interrelations between

E. Macioszek and G. Sierpiński (Eds.): TSTP 2021, LNITI, pp. 79–88, 2022.
https://doi.org/10.1007/978-3-030-93370-8_6

innovative processes and economy sector development stages. Also, particular attention is paid to effective distribution and sales of products [2, 3, 7–9].

2 Materials and Methods

The knowledge is critical at all stages of the industry development life cycle. Based on a data base that identifies innovative activity for individual states and specific industries for the United States, the empirical evidence suggests that the propensity for innovative activity is shaped by the stage of the industry life cycle [3]. New knowledge is generated through innovative activities, whose clustering at early life cycle stages and a subsequent dispersion at the maturity and decline stages are predefined by a degree of geographical concentration of the place of production. This may suggest that the positive agglomeration effects during the early stages of the industry life cycle become replaced by congestion effects during the latter stages of the industry life cycle [3].

The study [7] identified a key asymmetry, arising out of the disposition of problems against the main problem: larger problems in component innovations increase benefits received by technology leaders, whilst more serious problems in additional component innovations offset these benefits. The effect of a vertical integration as a strategy for managing ecosystem interdependence tends to grow over the technology life cycle.

In historical perspective, the move from a linear innovative process organization model towards a nonlinear one is obvious [6, 10, 11]. The linear model is a plurality of successive stages, where scientific research is the only source of innovations. G. Mensch [12] in his study identified two main aspects of the innovative process development: technological impetus is the basis of innovative changes, and the depression is a trigger of innovative activities. Another kind of a linear model was substantiated by K. Freeman [13, 14], and other researchers. According to K. Freeman [13, 14], the development of innovations is stimulated by rising demand, which, in its turn, initiates a diffusion of product & process innovations. The theory of K. Freeman [13, 14] and his followers is called the "demand pressure hypothesis". It was the "demand pressure" that significantly intensified innovative processes in the early 1980s.

A significant difference between the Russian and foreign approaches is, first of all, that the latter took into account consumer and market demands and involved obligatory marketing research at the final stage of the innovative process during sales and promotion of new products to the markets.

In this regard, a comprehensive estimate of life cycle costs should be based on a non-linear innovative process organization model and cover all process stages, from the birth of a productive idea to the product withdrawal from the market. In the developed economies, all elements of the innovative process, such as scientific studies, design, research and process developments, investment and financial resources, institutions and tools, marketing activities, production capacities and organizational structures, integrated in one piece, are focused on one goal, i.e., on the creation and promotion of innovations. At the same time, arranging feedbacks based on the compliance with specified parameters of constituent life cycle elements allows the responsibility of each subject of the innovative process to be determined.

In the study [4], the authors, having evaluated a parametric duration model, concluded that companies benefit from a premium for innovations, which extend their life cycle regardless of such indicators as age and size. Chances of the company survival tend to rise with the company age and growth rate, with the latter being more important than the original size of companies.

The firm size distribution changes significantly as an industry goes through stages of its life-cycle. The evolutions of the employment and output distributions also differ significantly, but display strong inter-industry regularities, including that the nature of the evolution depends on whether the industry is experiencing growth, shakeout, maturity, or decline [8]. Key advantages of using the innovation life cycle theory in the innovative process management are as follows:

- time factor consideration,
- compliance with process-oriented approaches to management,
- process central trend identification,
- visible conversion and transformation dynamics,
- logical process evolution,
- transparency of inventory, information and financial flows,
- mathematical modeling of stages and processes,
- identification of interrelations among various economic objects.

Key innovation activity vectors in the railway transport are as follows [9]: improve transport process management system; provide a harmonized and well-balanced transport infrastructure development; renew and upgrade rolling stock; enhance train traffic and safety management systems; increase operating reliability and extend running resource of equipment; implement energy- and resource-saving technologies; develop fast and high-speed rail transport; improve quality of transport services; increase cost efficiency of core activities; reduce burden on the environment; improve environmental safety.

It's worth mentioning that numerous studies confirm the need for implementing innovative activities at all company development stages. For example, the authors in the study [4] concluded that the high-tech sectors provide the most favorable opportunities for a company's survival [4].

Innovative activities, resulting in the product or process innovations [15] and which are based on the life cycle theory [6], are considered as a process, dynamically synchronized in time and space, for providing resources and defining liabilities of each participant for sci-tech support and development, production, promotion and implementation of innovations. In this regard, an emerging innovation management system of the company should follow a process-oriented approach to the company's innovation policy implementation. Process management, being the resultant of achievements in individual business processes, is focused on the overall result and thus decreases the number of hierarchical levels in management, and also facilitates an accurate determination of both overall and local results.

While justifying development programs, the implementation of the company's comprehensive innovation policy based on the life cycle theory allows taking into account requirements for life cycle cost parameters of technical systems to ensure a feedback between innovative process elements and the demand for transport services.

Typically, initial costs of innovative technical systems in the railway transport are higher as compared to traditional ones. However, at the same time, they must provide lower operating costs in comparison to the existing analogues [6]. In this regard, the requirements to be developed for promising technical systems should contain a clause requiring developers, manufacturers and suppliers to provide information on parameters of the compliance of quality and reliability indicators of industrial high-tech products with the product life cycle cost estimates.

Life cycle costs of a technical system include the consumer costs associated with the purchase and ownership of the system, i.e. the purchase price, associated lump-sum expenditures, operating costs, with the consideration of after sales services over the entire life cycle, and disposal costs. It should be emphasized that these costs should also be included in the life cycle costs while making accompanying investments.

During a technical assignment development for a new technical system, suppliers should estimate its maintenance and repair costs. The customer's functional responsibilities include estimates of the equipment life cycle costs and the analysis of the "sensitivity" of the life cycle cost parameters to changes in various components and individual elements. Life cycle cost models used by suppliers and end users can differ both in terms of the number of parameters and the quality of input data. Results obtained within the frames of the life cycle cost concept are a tool for justifying decisions on ordering and purchasing the technical system.

The analysis of methodological approaches to a technical system life cycle cost estimate showed that these approaches are based on the estimates of lump-sum costs (investments) and current costs (operating expenditures) over the useful service life of the system. The difference in approaches is associated with the consideration of disposal costs, warranty service costs, unscheduled repair costs, downtime costs, costs of additional operations, etc.

For the purpose of modeling, a life cycle of technical systems is presented as a process consisting of six main stages: concept definition and technical assignment development; research and development; technical system manufacture; implementation; operation and maintenance; dismantling and disposal.

At the concept definition and technical assignment development stage, the customer and the supplier of a technical system perform marketing studies, formulate initial technical requirements for the system, arrange a competition (tender) among suppliers for development and manufacture, initially estimate expected life cycle costs of a piece of the technical system, and develop basic provisions of a draft agreement for system development. The outcome of the work completed at this stage is to decide on the possibility and feasibility of the technical system development in terms of the consumer and commercial parameters, develop and approve technical requirements for a specific type (series), and select a technical system supplier.

At the research and development stage, R&D activities are performed with the aim to find the ways of and principles for a rational development of a new technical system, develop technical assignments, estimate in detail life cycle costs of the technical system, and proceed with R&D for prototype development, production and testing. The outcome of the work completed at the R&D stage is to produce a technical system prototype and to approve design documentation.

At the manufacturing stage, preparatory processes are carried out to ensure the enterprise readiness for production and output (delivery) of the product in a specified quantity and in accordance with the technical requirements, technical assignment, design documentation, technical specifications; also, manufacturing processes are launched as the established production line. The outcome of the work completed at the manufacturing stage is to produce a sample of the new technical system.

At the implementation stage, the equipment is brought into operation and accompanying measures for personnel training, additional repair base furnishing, etc., are taken. The outcome of the work completed at this stage is to obtain the assembled and ready-to-use technical system.

At the operation stage, the customer's operating organization (subdivision) accepts the technical system, brings it into the operating fleet for direct use in accordance with the intended purpose (in particular, during the warranty period), and maintains the specified availability level of the technical system fleet by implementing a set of measures (including maintenance and repair) with the aim to ensure and/or restore their operability and serviceability. The outcome of the work completed at the operation stage is to proper operate the technical system and to assess the compliance of actual reliability indicators and cost parameters with the specified ones.

The disposal stage anticipates the implementation of a set of documented organizational and process-related measures to write off, decommission and dismantle the technical system and to dispose and remove waste from its components. The outcome of the work completed at this stage is to dismantle a set of parts for reuse during repair, or destroy worn-out components, which are unsuitable for further rehabilitation.

Considering the above, life cycle costs of a technical system includes costs associated with the purchase (stages I-IV), ownership (stage V) and disposal (stage VI). Life cycle costs of the railway transport technical systems are defined by summing up the individual outflow of funds at each time interval of the life cycle.

Substantiation of decisions on purchasing railway transport technical systems based on the life cycle cost analysis requires establishing a standardization system and allocating costs to the prime costs by type of activity as well as monitoring and controlling thereof at separate phases and stages. The analysis of causes of discrepancies between actual values and design life cycle cost parameters enables the formulation of technical requirements for new elements of the railway transport technical systems, as well as the determination of economic entities' liabilities at individual phases and stages.

Life cycle costs of the railway transport technical systems, taking into account suppliers' liabilities for a failure to provide compliance with the technical system parameters [6], are suggested to be determined by the following formula:

$$LCC = PP + \sum_{t=1}^{T} (OPEX_t + ALC_t - DV_t) \cdot \alpha_t + \sum_{t=1}^{T} (CUR_R - CP_t) \cdot \alpha_t \quad (1)$$

where:

PP - technical system purchase price, thousand rubles,
$OPEX_t$ - annual operating and maintenance costs of the technical system, thousand rubles,
ALC_t - accompanying lump-sum costs associated with technical system commissioning, thousand rubles,
DV_t - disposal value of the facility, thousand rubles,
α_t - discount factor,
t - current year of operation,
T - useful life of the technical system, years,
CUR_t - cost of unscheduled repairs of technical systems caused by the fault of the supplier (developer, manufacturer), thousand rubles,
CP_t - supplier compensatory payments for a failure to provide compliance with life cycle parameters, thousand rubles.

To assess the impact of the technical system reliability indicators on its life cycle costs, the purpose, operating conditions, mode of operation and reparability of the system should be taken into account. Reliability indicators are divided into five main groups: reliability indicators, durability indicators, reparability indicators, preservation indicators and integrated technical system reliability indicators. Reliability indicators should be selected with due consideration of the conditions of use of technical systems, by applying the following principles:

- if, according to the conditions of use, the technical system is supposed to be repaired during operation, then the availability factor or the technical utilization factor (integrated indicators) should be selected as one of the reliability indicators,
- if a failure of a technical system or its individual components results in a failure of the system to perform its critical process tasks, violates safety of the railway transport operation or causes a threat to human health and life, then fail-safety indicators, expressed as the mean time between failures (MTBF), or the trouble-free operation probability, should be the main reliability indicators of these systems,
- if a technical system downtime due to a failure incurs large material costs, then such a system should be featured with an appropriate reparability and high failure-free operation indicators as well as minimum average time of post-failure recovery.

3 Results

To determine the probability of failures or confidence intervals of the assessed reliability indicators of technical systems, statistical methods are applied. Key vectors of monitoring and controlling reliability indicators of technical systems are as follows:

- timely verify reliability indicators and validate their compliance with standard values of life cycle cost parameters of technical systems,
- analyze the main causes of potential critical problems in the technical system operation with the aim to prevent failures and deviations,

- develop recommendations for the implementation of cost-effective preventive and corrective actions based on the risk analysis of non-compliance of life cycle cost parameters,
- control processes used for ensuring accuracy and completeness of information required for verification of reliability indicators, and assess their impact on life cycle costs.

Costs, associated with the fulfillment of specified requirements for reliability indicators of technical systems, are classified as follows: costs incurred due to a failure to provide compliance with specified reliability indicators; technical system recovery costs, including repair costs; maintenance costs; and additional costs. Figure 1 schematically shows the impact of reliability indicators on operating and repair costs of technical systems.

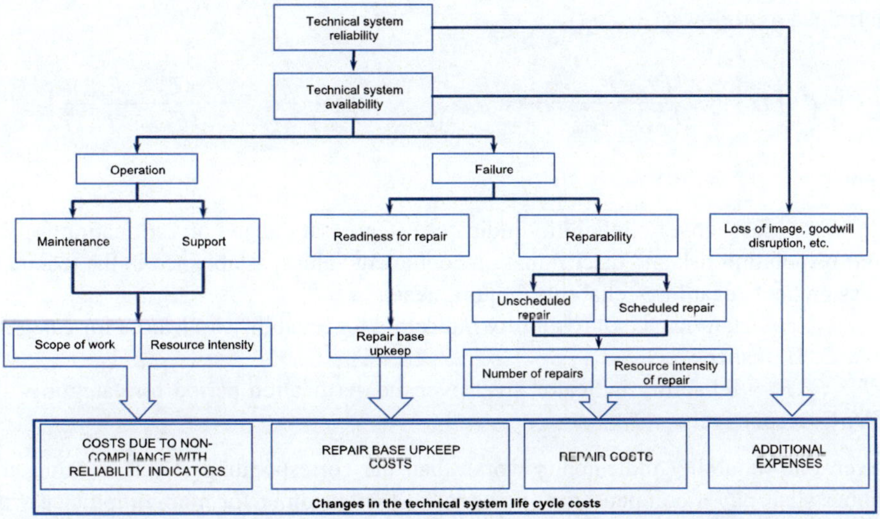

Fig. 1. Impact of reliability indicators on the technical system life cycle costs

The compliance of reliability indicators with regulatory and technical documentation requirements is confirmed by an experimental determination of quantitative and qualitative reliability characteristics of technical systems during their operation, modeling or other impacts. If the required level of reliability indicators, specified in the contract, is not achieved within the basic and extended verification periods, the supplier shall pay a compensatory payment calculated according to the following formula:

$$CP = r \cdot P \cdot \frac{(F - F_{\min})}{F_{\max} - F_{\min}} \tag{2}$$

where:

r - percentage of compensation for non-compliance with reliability indicators of the total contract price [%],
P - price of a verified batch of technical systems (contract price), [rubles],
F - actual reliability indicator level [ea],
F_{max} - reliability indicator level adjusted for the excess factor specified in the contract [ea],
F_{min} - permissible reliability indicator level specified in the contract [ea].

Improved reliability indicators in one group, even if comparable in percentage terms, does not always compensate for deterioration in reliability indicators in another, more significant group. Enhanced reliability entails the reduction in unplanned repair costs throughout the life cycle. Compensatory payments, provided that the worsening of failures in one category is not offset by improved indicators of another category, are determined as follows:

$$CP = P \cdot \left(r_A \cdot \gamma_A \cdot \frac{\left(F^A - F^A_{\min}\right)}{\left(F^A_{\max} - F^A_{\min}\right)} + r_B \cdot \gamma_B \cdot \frac{\left(F^B - F^B_{\min}\right)}{\left(F^B_{\max} - F^B_{\min}\right)} + r_C \cdot \gamma_C \cdot \frac{\left(F^C - F^C_{\min}\right)}{\left(F^C_{\max} - F^C_{\min}\right)} \right) \quad (3)$$

where:

A, B, C - respectively, reliability indicator category at the end of verification for the corresponding risk and discrimination coefficient values, established in the technical assignment regarding reliability requirements,
$\gamma_A, \gamma_B, \gamma_C$ - significance of reliability indicators for reliability indicators for category A, B, C, respectively,
F^A, F^B,F^C - reliability indicator level over the verification period for categories A, B, C, respectively.

If even one reliability indicator is worse than the corresponding standard value, the supplier shall pay a compensatory payment to the customer for non-compliance with life cycle parameters. The reliability indicator, with the consideration of the compensatory payment for the deterioration of failures in one category by improving indicators in another category, is calculated by the formula:

$$F = F^A \cdot \gamma_A + F^B \cdot \gamma_B + F^C \cdot \gamma_C \quad (4)$$

The results of using the proposed methodological approach to assessing the size of compensatory payments for non-compliance with life cycle parameters, at a five percent compensation value (r_κ), are presented in Table 1.

The importance of the category of failures in the assessment of economic responsibility is stipulated in the contract and for the given calculation it is accepted at the level of average values set by the JSC "Russian Railways". The normative rate of failures is chosen according to the test plan drawn up in accordance with the international standard of "IEC 61124:2012 "Reliability testing - Compliance test for constant failure rate and constant failure intensity". The actual rate of failures is determined on the basis of the JSC "Russian Railways" statistics on the failures of technical systems in terms of failure categories. The data indicated in Table 1 show that if the deterioration of failures in one category is not offset by improved indicators in another category, the supplier's liability is RUB 8.6 million; in case if the deterioration of failures in one category is offset by improved indicators in another category, it will be RUB 9.2 million.

Table 1. Assessment of the size of compensatory payments for non-compliance with life cycle parameters.

Indicator	Failure category			Integral score
	A	B	C	
Significance of the failure category in assessing liabilities, frac	0.5	0.3	0.2	1.0
Standard failure rate, ea	4	22	66	21.8
Actual failure rate, ea	3	26	70	23.3
Minimum failure rate*, ea	3	21	64	20.6
Maximum failure rate*, ea	5	23	68	23.0
Failure deviation assessment for liability assessment, ea	0.0	2.5	1.5	1.13
Compensatory payment for non-compliance with life cycle parameters, RUB mln	0.0	6.1	2.5	9.2

Note: *minimum and maximum failure rates with a breakdown by category are indicated in the equipment supply contract

In our opinion, contracts for the development and purchase of sophisticated technical systems for railway transport should include methods for verification of key parameters embedded in the life cycle cost estimates, as well as an adequate liability scale. Using liability claim mechanisms, transport companies will be able to minimize risks, associated with a significant increase in unscheduled repair costs due to a failure to provide reliability indicators of supplied technical systems. Thus, a comprehensive estimate of life cycle costs of technical systems allows an objective assessment of the impact of reliability indicators while justifying management decisions as well as the identification of parties responsible for a failure of developers and suppliers of innovative industrial products to deliver key parameters.

4 Conclusions

Innovative activity is the main source of increasing transport company competitiveness through the implementation of economic benefits of innovations in comparison with existing technical systems. To intensify innovative activities, non-linear innovative process organization models should be used to expedite the implementation of scientific achievements into practice.

The "life cycle cost" indicator, enabling the identification of competitive advantages of a new-implemented technology in comparison with existing analogues, should be used as a criterion for justifying decisions on the new technical system implementation.

Verification of reliability indicators helps to determine the liability of developers, manufacturers and suppliers for a failure to provide compliance with life cycle cost parameters. In this case, the impact of technical system reliability indicators on its life cycle costs should be taken into account.

Using the above mentioned methods for estimating the liability of suppliers of technical systems for railway transport for non-compliance with declared life cycle cost parameters will improve the accuracy of life cycle cost estimates and the relevance of management decisions on the system upgrade.

Comprehensive estimate of life cycle costs of science-intensive and technically sophisticated railway transport products can be used as the basis of a professional liability insurance system defining the liability of developers, manufacturers and suppliers for compliance with specified reliability indicator levels.

References

1. Ameri, F., Dutta, D.: Product lifecycle management: closing the knowledge loops. Comput. Aided Des. Appl. **2**(5), 577–590 (2005)
2. Agarwal, R., Gort, M.: Firm and product life cycles and firm survival. Am. Econ. Rev. **92**(2), 184–190 (2002)
3. Audretsch, D.B., Feldman, M.P.: Innovative clusters and the industry life cycle review of industrial organization. Rev. Ind. Organ. **11**(2), 253–273 (1996)
4. Cefis, E., Marsili, O.: A matter of life and death: Innovation and Firm Survival. Ind. Corp. Chang. **14**(6), 1167–1192 (2005)
5. Tereshina, N.P., Podsorin, V.A.: Economics of Innovations in Transport. Russian University of Transport, Moscow (2019)
6. Tereshina, N.P., Podsorin, V.A.: Life Cycle Management of Technical Systems in Railway Transport. Vega-Info, Moscow (2012)
7. Adner, R., Kapoor, R.: Value creation in innovation ecosystems: how the structure of technological interdependence affects firm performance in new technology generations. Strateg. Manag. J. **31**(3), 306–333 (2010)
8. Dinlersoz, E.M., MacDonald, G.: The industry life-cycle of the size distribution of firms. Rev. Econ. Dyn. **12**(4), 648–667 (2009)
9. Tereshina, N.P., Podsorin, V.A.: Innovation Management in Rail Transport. Vega-Info, Moscow (2012)
10. Schumpeter, J.A.: The Theory of Economic Development. Direct-Media, Moscow (2007)
11. Kondratyev, N.D.: Big Conjuncture Cycles. Urait Publishing House, Moscow (2000)
12. Mensch, G., Continho, C., Kaasch, K.: Changing capital values and the prosperity to innovate. Futures **13**(4), 276–292 (2019)
13. Freeman, C.: Technical innovation, diffusion, and long cycles of economic development. In: Vasko, T. (ed.) The Long-Wave Debate. Springer, Berlin, Heidelberg (1987). https://doi.org/10.1007/978-3-662-10351-7_21
14. Freeman, C.: Technical innovation, diffusion, and long cycles of economic development. In: Vasko, T. (ed.) The Long-Wave Debate, pp. 295–309. Springer, Berlin, Heidelberg (1987). https://doi.org/10.1007/978-3-662-10351-7_21
15. Podsorin, V.A.: Economic Assessment of a Transport Company Capitalization. Russian University of Transport MIIT, Moscow (2007)

Development of Technology for the Formation of Innovative Logistics Products - Components of the Transport Holding Ecosystem

Olga V. Efimova, Elena Z. Makeeva, Irina G. Matveeva(✉), Vitaliy M. Morgunov, and Ekaterina N. Mironova

Russian University of Transport, Moscow, Russia
matveeva07@list.ru

Abstract. One of the ways to increase the efficiency of the transportation process and the development of transport and logistics services of railway transport is the use of modern design tools and economic justification of new logistics products of the ecosystem of the Russian Railways holding. In the context of the need to increase the production capacity of infrastructure and the capacity of railways, as well as striving for the fullest satisfaction of customers' expectations of high quality services, supplemented by special conditions and individual solutions, the transport holding is tasked with operational design and development of new logistics products. The article discusses the concept of transport and logistics products using the best practices of the product framework.

Keywords: End-to-end processes · Digital environment · Innovative logistics products · Transport holding ecosystems · Product framework

1 Introduction

The evolution of operational excellence in all areas of business is progressing from eliminating waste and engaging staff in a culture of continuous improvement to scaling "quick wins" into innovative, competitive products. In the 20s of the 21st century, one of the leading directions for the development of business models is the formation of cross-functional end-to-end processes that combine disparate operations into a value chain using digital technologies. This approach forms an innovative product portfolio, enhancing the role of the product owner in the business and in his interactions with customers.

The attractiveness and demand for transport and logistics services of JSC Russian Railways directly depends on the degree of synergy between the functional branches and subsidiaries and dependent companies (SDCs), which form end-to-end innovative solutions for the digital pricing environment in the ecosystem of the Russian Railways holding company, modern payment methods for work performed and services rendered [1, 2].

E. Macioszek and G. Sierpiński (Eds.): TSTP 2021, LNITI, pp. 89–100, 2022.
https://doi.org/10.1007/978-3-030-93370-8_7

An important modern vector for the formation of adaptive transport and logistics products is digital approaches to the development of business initiatives aimed at reducing losses in production processes based on data analytics.

In the context of the formation of a single ecosystem of a transport holding, which includes an extensive network of services, services and their suppliers linked by a single digital platform, it is necessary to solve the problem of finding optimal approaches to a new pricing model for innovative transport and logistics products [3].

The use of blockchain-based smart contracts technology allows creating a transparent information environment (distributed data register) for all transportation participants with objective recording of the history of changes in the state of the cargo in the processes of loading and transportation. The principles of this solution can be applied in practice in the transport and logistics activities of JSC Russian Railways. The Smart Contract technology fixes the conditions and obligations of each participant, makes the responsibility and progress of these obligations "transparent", on the basis of which the payment for transport and logistics services is automatically carried out.

In the complex of scientific and methodological problems of the formation of innovative logistics products, it should be noted the need for:

- implementation of a product approach in order to accelerate the implementation of successful business projects,
- modification of price instruments for the sale of logistics products,
- development of conceptual approaches to business stability in conditions of information uncertainty,
- implementation of smart contract technology,
- building a model for assessing the effectiveness of innovative transport and logistics services.

New logistics products and business initiatives should be preferentially aligned not only with business sustainability goals, but also with a broader range of sustainable development goals and their ethical and strategic norms, which requires analysis and audit of new projects to achieve carbon neutrality targets and sustainable development.

2 Materials and Methods

The product approach aimed to create an environment to accelerate the delivery of successful product initiatives. The introduction of a product approach allows us to speed up the creation and implementation of new products by 2–2.5 times, to ensure the traceability of implemented initiatives, to increase the number of ideas for each monetary or natural product meter. The best practice for introducing a product approach is to create a unified system for the development of management and control over the activities of product teams - the "product framework". The development of the status and functionality of product owners with target Time to Market forms an organizational model of a product laboratory, which, in fact, becomes a Product Management Office, which, manages a product portfolio with a target PNL (profit and loss report), identifies efficiency blockers.

The introduction of this approach in the transport sector requires the development of a technology for the formation of innovative logistics products, which will include tools for assessing and selecting innovative logistics products and business models for their implementation in a transport holding based on economic criteria, taking into account sustainable development goals. The introduction of the "product framework" system is due to the need to substantiate and confirm the calculated values of natural and cost performance indicators, as well as effects, including managerial ones, from the implementation of new transport and logistics services aimed at improving the quality of services provided to the clients of the transport company, as well as innovative solutions pricing and payment methods for the provided transport and logistics services. The pricing model operating in railway transport with a variety of tariff schemes depending on the type of rolling stock used does not correspond to market conditions and leads to the "losses" of high-tech goods in favor of alternative modes of transport. Creating a digital environment and a new pricing model involves:

- taking into account the economic parameters of a specific transportation,
- an individual approach to pricing for each client,
- adaptability of pricing to market fluctuations,
- the formation of an economically sound digital pricing environment based on the principles of end-to-end planning, taking into account exogenous and endogenous factors, is based on a comprehensive analysis of modern methodological approaches to the pricing of innovative transport and logistics products in the transport market in terms of their competitive advantages, as well as on the principles of marginal pricing,
- formation of an information and analytical base for the development of solutions in the field of pricing of innovative logistics products using digital technologies, clustering of goods by the transport component in the price of transported goods, taking into account the dynamics of market conditions in the long term.

The target vision of the Smart Contract service in the transport and logistics activities of JSC Russian Railways is formed taking into account the modern requirements of the transport market in creating a single information space for all participants in the transportation process, including: consignor and consignee, rolling stock operators, payers, owners of non-common railways use, federal government bodies, seaports. That is, the service will create a "digital trust environment" that ensures the transparency of the transportation process with the automatic fulfillment of contractual and financial obligations, will lead to the optimization of document flow and claims work, and will reduce operational and time costs for all participants in the transportation.

At the same time, the service will simplify the existing business processes in the organization of cargo transportation, provide contractual support for transport and logistics activities, provide all participants, depending on their roles, the opportunity to receive timely information on the execution of the contract, including financial settlements, while ensuring the unconditional reliability of the data. reflected on the blockchain platform.

Speaking about the relevance of this issue of forming an ecosystem of tariff solutions that should ensure the sustainable development of the transport market, it should be noted that complex modern systems should primarily rely on digital services. Going beyond the consideration of this issue for a particular mode of transport and forming transport products based on end-to-end services, it is necessary to revise the mechanisms of tariff setting for various modes of transport in the direction of their harmonization. The modern mechanism of tariff solutions, taken within the framework of price limits, taking into account international experience in the establishment and application of price mechanisms, is aimed at providing users of railway transport services with tariff preferences. The development of digital transport and logistics products and services of the Russian Railways holding company, with the participation of partners-major players in Russian and foreign business, requires the formation of tools for assessing the effectiveness of new transport and logistics services. The main task of the formation of an ecosystem is a wide coverage of clients with a full range of diverse and integrated services [4]. Economic substantiation of innovative logistics products must provide a list of calculation that take into account various conditions for the implementation of the project: attraction of innovative rolling stock (purchase, leasing, rent); application of various models of operation of technical means (responsibility of a functional branch, sourcing); implementation of modern information systems that support the activities of the transport and logistics business unit of the Russian Railways SDC holding company (purchase of licenses, software development, maintenance, etc.) and other possible options for project implementation. To form a business model, a list of assumptions and assumptions must be defined.

All cash flows must be calculated separately for subsidiaries and dependent companies (if they are implementing the project) and for JSC Russian Railways. For JSC Russian Railways flows should be calculated "project based" - for certain business models and "without a project" - the implementation of the existing technological process without changes. Information uncertainty in the development of commodity markets gives rise to internal crises in transport companies, when some business units become redundant, external crises, when the customer base is massively changing, and in some cases - a double crisis, internal and external at the same time, threatening the existence of a transport holding. These crises require different strategic and tactical approaches. It should be borne in mind that a rapid response strategy with high information uncertainty is dangerous. With an external crisis, the organizational transformations of the business contain a danger, including a change in structure that can lead to a breakdown in personal ties, which are essential in the transport business.

The Fig. 1 shows the concept of strategic approaches to information uncertainty in the transport and logistics business. In the absence of regulations and agreements between the participants in the implementation of cross-functional transport and logistics products, with the growth of information uncertainty, the speed of adaptation to changing conditions increases (straight line). However, business utility (value-V) in these conditions is small, and the costs of adaptation to significantly changing conditions will require significant (C_v), which can lead to chaos and irrational interaction of participants in a cross-functional product.

In the presence of clearly defined rules and regulations of interaction, the speed of adaptation to changing conditions is very low. However, the predictability of the actions of the participants in a cross-functional product is quite high (dashed line), the costs of achieving concerted actions (C) are insignificant, and the resulting utility (V_c)

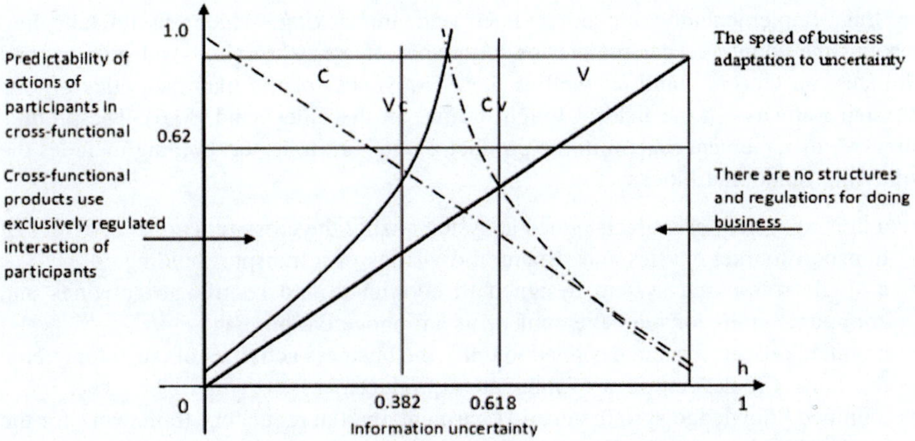

Fig. 1. Conceptual approaches to business stability in conditions of information uncertainty

is determined. There is a zone in which, under conditions of uncertainty according to the principle of maximin in decision-making, the best of the worst results of cross-functional interaction can be selected. It is formed around the intersection of straight lines, the speed of adaptation and the predictability of the actions of participants in innovative transport and logistics products. Under these conditions, the VUCA rules (Volutility-variability, Uncertainty-uncertainty, Complexity-complexity, Ambiguity-obscurity) of management apply: the cancellation of past experience and excessive detailing are harmful, immediate adaptation to each change is impossible, new ideas and a generalized view of efficiency are needed, constant search for additional information.

The presence of an external and internal crisis, as well as their joint manifestation, requires from the transport business not only adaptation to information uncertainty, but also an internal transformation of the business model, which provides greater freedom and avoiding the regulation of internal actions of employees, and especially experts who interact with clients.

3 Results

The development of new logistics products and services is carried out in various divisions of the transport and logistics block of the Russian Railways transport holding. Currently, technologies for designing innovative logistics products and building business models for their implementation are not defined and are lacking, which should take into account the economic parameters for the owner of a new service - subsidiaries and affiliates, on the one hand, as well as the effectiveness of the main activities of Russian Railways in the field of freight traffic, on the other hand [5, 6]. The development of new logistics products and services is carried out in various divisions of the transport and logistics block of the Russian Railways transport holding. Currently, technologies for designing innovative logistics products and building business models for their implementation are not defined and are lacking, which should take into account the economic parameters for the owner of a new service - subsidiaries and affiliates, on the one hand, as well as the effectiveness of the main activities of JSC Russian Railways in the field of freight traffic, on the other hand [5, 6]. The product initiative management system in the product framework transport holding includes the following functional blocks:

- a unified system of artifacts aimed at systematizing the cross-functional activities of transport market entities and structural divisions of a transport holding,
- a single role-based system designed to coordinate and coordinate activities and communications for the development of an innovative product,
- a unified process system designed to unify the business activities of cross-functional teams for the development of innovative products,
- a unified knowledge system aimed at harmonizing the regulatory framework for the implementation of activities in the framework of the implementation of innovative products,
- a unified system for assessing the effectiveness of activities and operations in the framework of providing an innovative product to the transport services market.

An important technique of the product framework, which is used in the practice of companies operating in the market of software products, is the formation of a user flow. With regard to the transport and logistics market, this is of particular importance for facilitating interaction with participants in cross-functional products. This map shows the main stages of the implementation of a transport and logistics service, the participants in its implementation and the possibilities of an innovative product in terms of eliminating customer losses. This map will also undergo transformation after the launch of the MVP after receiving feedback from the first customers. They will identify opportunities for product improvement and these transformations will need to be reflected on the map. Thus, the first stage of creating a viable innovative transport and logistics product based on the MVP concept includes the following steps:

- defining the basic principles of creating an MVP,
- finding the problem of customers and participants of cross-functional interaction, which will be solved by the MVP,
- search for target audience,
- identification and analysis of the main competitors,
- conducting a SWOT analysis,
- creation of a customer journey map,
- drawing up a list of product functions,
- definition of an extended list of MVP functions,
- testing.

The main tasks solved within introducing blockchain technology into innovative logistics products this is:

- automation of the function of concluding an automatically executed electronic contract,

- automated recording of each technological stage of transportation (contract parameter) in the distributed data register with verification of its compliance with the standards of operations and the terms of the contract,
- automatic recognition by the parties of the agreement of responsibility in case of its violation (absence of claim procedures) and carrying out mutual settlements,
- and finally, a phased transition to a single electronic form of contracting for the provision of a comprehensive transportation service and an innovative logistics product.

For legal and contractual support of the service block chain a "form of a framework agreement on the self-fulfillment of the contract of carriage and related contracts for the use of blockchain technology" was developed, which included production rules (textual description of the algorithm and options and/or non-fulfillment of contractual obligations) and the regulation of mutual settlements.

For the period 2020–2021, a list of technological operations has been determined, such as:

- the procedure for connecting external users and the cost of providing access and maintenance of the DRG FT was calculated (distributed data register "Freight transport"),
- an algorithm has been created that takes into account the fulfillment of contractual and financial obligations,
- the number of data sources has been increased.

When comparing the target vision and analyzing the current situation, a number of problems were identified, due to which a large-scale fully functional implementation of the project cannot proceed at an accelerated pace. Firstly, this is due to the technological features of the project implementation, since much of the line-level processes are not fully automatic. This problem must be solved through phased automation. Another tangible challenge is the legal framework for smart contract technology. At the same time, the possibility of using a smart contract as a form of agreement and as a mechanism for fulfilling an obligation is provided for by the legislation of the Russian Federation. Despite the risks and difficulties in the implementation of the project, we understand the importance and necessity of achieving the target state. This is necessary primarily for clients due to the reduction in the time of operations and the guaranteed fulfillment of contractual obligations.

To substantiate the feasibility of introducing innovative logistics products into the transport business, a model is being formed, which includes the following flows with the distribution by years of the implementation of the project for the development and launch of a new innovative product on the market:

- investment cash flow (for example, investments in the purchase of rolling stock, the arrangement of logistics terminals, the purchase of loading and unloading equipment, vehicles, the development of information systems, the creation of a project office), etc.,
- cash inflows (for example, payment for transportation, rental of rolling stock, income from additional services (storage, loading, insurance, security), income

from the delivery of goods by vehicles owned by subsidiaries and affiliates (in the case of providing a comprehensive service), etc.,

- cash outflow, taking into account the operating costs of the project (for example, the tariff for the carriage of goods, repair of rolling stock, maintenance of information resources, etc.) without depreciation and tax payments.

Based on the above flows, a cash flow should be formed, which is discounted taking into account the accepted discount rate. The model should make it possible to determine the feasibility of implementing an innovative logistics product in a transport holding based on the calculated NPV (net present value), IRR (internal rate of return) and the discounted payback period. Studies have shown that many large companies, when developing innovative logistics products, adhere to the concept of sustainable development and, in their development plans, include measures for resource conservation, energy conservation, environmental protection and prevention of man-made disasters. JSC Russian Railways in its activities gives priority to the principles of sustainable development both in relation to social programs for its employees and in relation to environmental issues and environmental protection. The activities of JSC Russian Railways in the field of climate change mitigation are based on the provisions of the Environmental Strategy of JSC Russian Railways for the period up to 2020 and for the long term until 2030 and the Energy Strategy of JSC Russian Railways for the period until 2020 and for the future period until 2030, as well as the Climate Doctrine of the Russian Federation.JSC Russian Railways through the introduction of new technologies and contributes to the achievement of the national goal - to reduce the volume of greenhouse gas emissions to a level of no more than 75% from 1990.

Management of greenhouse gas emissions is a part of the implemented corporate strategy of Russian Railways. This allows the Company to contribute to the achievement of the national goal - to ensure the reduction of greenhouse gas emissions to a level of no more than 75% from 1990. The reduction will be achieved due to new technologies that reduce fuel consumption in traction and stationary energy. In particular, in traction power, when autonomous diesel heating systems are used on diesel locomotives, fuel consumption is reduced by 10–16% during the idle time of diesel locomotives in a "hot" state. The introduction of modular gas-fired boiler houses instead of coal and fuel oil-fired boiler houses significantly reduces the number of emissions of harmful substances into the atmosphere. As part of the implementation of resource-saving technologies, projects for equipping railway stations with resource-saving technologies are being implemented within the framework of the adopted Smart Station concept, the introduction of LED lighting, automated central heating points, energy-optimal timetables for passenger and freight trains, and automatic driving systems on locomotives. The transition to new types of rolling stock plays an important role in reducing greenhouse gas emissions. In particular, the indicative calculation of the impact on the environment as a result of the use of electric trains "Lastochka" showed a significant advantage of rail transport for the carriage of passengers in comparison with road transport. For all new logistics projects, CSR reporting should also be prepared in accordance with the rules of the Russian Union of Industrialists and Entrepreneurs (RSPP).

4 Discussion

To improve the efficiency of a new product in the transport and logistics market, it is proposed to use the product framework method, which consists in offering the market a test version of a service or service with a minimum set of functions (sometimes even one) Minimal Viable Product, which has value for the end consumer. MVPs are created to test hypotheses and check the viability of the conceived product; how valuable and relevant it will be in the market. The results of testing the minimum viable product and feedback from the target audience help to understand whether it is worth developing the project further, what changes should be made to the strategy, and what should be left as is. The usefulness of MVP development is proven by examples of large companies. For example, a small, single-function music streaming service, launched in 2006, is now converted to a Spotify product and is valued at $ 21 billion, has partnerships with major recording studios, and has a 50 million active audience [7]. The analysis of the viability of an innovative product in the test version is complemented by a Proof of Concept (PoC). PoC describes the processes for ascertaining the technical viability of a software concept (or any other product), as well as describing the processes at the initial stage of the development of innovative products.

The most important thing about MVP is collecting information from early customers and shippers. The end consumer will tell about the correct model of project implementation or the disadvantages of the proposed one. The collected data can be used to plan further releases of the transport and logistics product model, taking into account the most priority goals of the customers of the transport market, as well as participants in cross-functional interaction: what functions to implement in the first place, what additional services it is advisable to expand the value of the proposed product with.

Vasilenko V.L. in his work "The main trends of digital logistics" indicates the problems of the development of the market for innovative transport and logistics products requires new tariff decisions, taken both within the price limits that meet the requirements of the current regulatory framework in the field of state tariff regulation, and ensuring an increase in the efficiency of key financial and economic indicators of railway transport and the development of the transport market [8]. The management of modern transport systems at any level cannot be imagined without taking into account the sustainable development factor. The emergence of this concept dates back to the 1980s and 1990s, when concern about the problems of irreversible changes in the environment ceased to be the lot of a narrow circle of researchers and acquired a global character [9]. The UN Sustainable Development Goals are aimed at taking measures to ensure the optimal use of limited resources and the use of nature, energy, and material saving technologies, maintaining the stability of social and cultural systems, the integrity of biological and physical natural systems. Sustainable development is such development that ensures the satisfaction of current needs without violating the interests of future generations. The World Commission on Environment and Development developed this definition in 1983. Failure to take into account the carbon footprint and objectives of achieving carbon neutrality carries the following risks for new projects:

- underfunding due to lack of access to green capital markets,
- insufficient quality of human capital in projects due to inadequate openness of the corporate and social reporting system,
- lack of priority access to government benefits and subsidies for technological development due to the lack of goal-setting in the logic of the SDGs.

Business models for new logistics products must reflect at least one of the ways to achieve carbon neutrality:

- reduction of direct emissions and transition to renewable energy sources - hydro generation, solar energy, wind energy,
- direct capture of $CO2$ from the air,
- compensation through investment in projects that reduce carbon dioxide emissions.

The priority area for the formation of innovative transport and logistics products is the business models of projects classified as "green". Green transport projects include: electric, hybrid, public, rail, non-motorized, multimodal transport, green vehicle infrastructure and emission reductions. In the works of O.V. Efimova discusses the issues of sustainable development of the Russian Railways company. Sustainable development is a qualitatively new stage in the evolution of a harmonious transport system, which is able to ensure effective interaction of transport market participants for economic growth, stimulate international trade, and reduce inequality within countries and between them [10]. It should be noted that these issues are regularly raised at international economic forums [9].

European logistics platform: Stability in the new reality is associated with the modern interpretation of high uncertainty as information incompleteness. For the transport and logistics business and its sustainable development, extensive and reliable information is required regarding the business environment of commodity producers and the dynamics of consumption, as well as places of predominant consumption of manufactured goods. Lack or excess of information, irrelevance of information, approximation of consumption dynamics in the retrospective period become risk factors for erroneous decisions in the field of product offers of the transport and logistics business [11].

5 Conclusions

The technology being developed for the design of innovative logistics products and business models for their implementation in the holding should take into account the functional and technological features of the new logistics products being created, which consist in changing the work technology, using new rolling stock, optimizing logistics traffic patterns, changing the functional responsibilities of employees, using sourcing technologies, in contrast to the implementation of investment projects (complex, projects for the development of railway transport, renewal and modernization of the rolling stock and infrastructure, social infrastructure of the company, etc.), which should ensure the implementation of the strategic objectives of the company's development by increasing and transforming tangible assets, comply with the general scheme for the development of the railway network.

The introduction of new cross-functional transport and logistics products, on the one hand, increases the business stability of the transport holding in the face of high volatility in the transport market environment and product markets, and on the other hand, with weak synchronization and unification of the regulatory environment, it can have a destabilizing impact on the transport business. Platform solutions for digital payment services for transportation of all types of transport, development of international digital transport corridors, and improvement of the regulatory framework for multilateral cooperation will allow further integration of tariff solutions within the Euro-Asian transport system until 2030. Transport companies providing rail transportation, transport, and logistics services with a predominant use of ecologically clean railway infrastructure, supports the principles of the UN Global Compact - the largest international initiative in the field of corporate social responsibility. The company is a member of the Association "National Network of Participants of the Global Compact for the Implementation of Responsible Business Principles into Business Practice". Along with the UN Global Compact, railway transport sector follows the principles of responsible and business practice enshrined in the Social Charter of Russian Business of the Russian Union of Industrialists and Entrepreneurs, regularly take part in the compilation of the collection of corporate practices of the Russian Union of Industrialists and Entrepreneurs. The commitment of Russian Railways to 17 UN goals in the field of sustainable development is revealed in the aspects of economic sustainability, environmental safety and social stability of the transport sector of the national economy of Russia.

Building a business model is an important part of the introduction of new transport and logistics products, in order to justify the feasibility of introducing which in the transport market, calculation and analytical tables should be developed, taking into account various factors for the implementation of the project: from the point of view of investment costs; the use of various types of operation of technical means; the use of new digital technologies that ensure the activities of the transport and logistics business unit of the Russian Railways holding company - the development of new software, the acquisition of licenses, support and other possible options for the implementation of projects. It is necessary to define a list of possible assumptions and assumptions in the formation of a business model.

References

1. The concept of development of the transport and logistics business of the Russian Railways holding. https://www.rzdlog.com/
2. Aksyonenko, N.E., Lapidus, B.M., Misharin, A.S.: Railways of Russia: from Reform to Reform. Transport, Moscow (2001)
3. Pustokhina, I.V.: Modern Trends in the Development of Logistics, pp. 339–345. Russian University of Economics. G.V. Plekhanov, Moscow (2017). https://www.rea.ru/en/Pages/default.aspx

4. Skrug, V.S.: Digital economy and logistics. In: Bulletin of BSTU im. V.G. Shukhov No. 5, pp. 138–143. Belgorod (2018)
5. Konotopsky, V.: Logistics: A Textbook for Universities, 4th edn. Yurayt Publishing House, Moscow (2020)
6. Negreeva, V.V., Vasilenok, V.L.: Logistics: Study Guide. ITMO University, Saint Petersburg (2015)
7. MVP: What it is and how it works. https://habr.com/ru/company/productstar/blog/508892/
8. Vasilenko, V.L.: The Main Trends in Digital Logistics. https://openbooks.itmo.ru/readeconomics/19478/19478.pdf
9. World Economic Forum: Digital Transformation of Industries. In: Logistics Industry, pp.1–31 (2016). https://reports.weforum.org/digital-transformation/wp-content/blogs.dir/94/mp/files/pages/files/wef-dti-logisticswhitepaper-final-january-2016.pdf
10. Vladimirovna Efimova, O., Zakharovna Makeeva, E., Igorevna Dmitrieva E.: Approaches to the formation of sustainable economic development of transport system. In: Proceedings of CBU in Economics and Business, vol. 1 (2020). https://ojs.cbuic.cz/index.php/peb/article/view/17
11. European Logistics Platform. http://www.european-logistics-platform.eu/en/members/

Safety Issues in Transport - Human Factor, Applicable Procedures, Modern Technology

Implementation of a Methodology for the Fatigue Analysis in Quay Crane Operators Through a Multi-agent - Discrete Event Simulation: The Salerno Container Terminal Case Study

Stefano de Luca, Lucas Joel Cisternas(✉), Chiara Fiori, and Roberta Di Pace

Transportation System Analysis Laboratory, Department of Civil Engineering, University of Salerno, Salerno, Italy
{sdeluca,lcisternas,cfiori,rdipace}@unisa.it

Abstract. Fatigue may have significant impacts on the port operations efficiency, but also on the safety of each single workers and on the "safety" of the whole terminal. To this aim, if on the one hand it is important to implement real-time solutions for improving port performances, on the other hand it is interesting to investigate the drop in performance of quay crane operators due to mental workload. The aim of this study was the simulation of the crane operator's fatigue. In particular, starting from real data provided on the two busiest work shifts, the paper aims at analyzing the crane operators fatigue curve in the port of Salerno and implementing it in a multi-agent discrete-event simulation model. The results of the analysis point out that the fatigue curve obtained through the real data referred to the port of Salerno are validated by the theory and consistent with the available fatigue curves.

Keywords: Crane operator fatigue · Quay crane performance curves · Fatigue · Terminal container simulation · Multi-agent discrete event simulation

1 Introduction

In the recent years, it has been observed a strong competition among Mediterranean ports which is mainly based on the operational efficiency of the logistic operations within the port and outside the port itself. Port efficiency may be achieved through an effective organization of the logistic process, using effective handling unit, and adopting advanced technologies. Nevertheless, the human factors continue, and will continue, to play a central role, and, in particular, the human fatigue due to the workload is a crucial issue which must be carefully considered and carefully managed in daily activities. Fatigue may have significant impacts on the port operations efficiency, but also on the safety of each single workers and on the "safety" of the whole terminal.

To this aim, if on the one hand it is important to implement real-time solutions for improving port performances, on the other hand it is interesting to investigate the drop

E. Macioszek and G. Sierpiński (Eds.): TSTP 2021, LNITI, pp. 103–119, 2022.
https://doi.org/10.1007/978-3-030-93370-8_8

in performance of quay crane operators due to mental workload. This issue leads to a decrease in the number of containers moved per shift, moreover may cause accidents that involve the crane operators as well as all the workers in the terminal and those who are working on the ships.

Currently, existing results highlight how quay crane operations are one of the main causes of unfortunate events and, at the same time, represent the most expensive insurance claim [1] for a port. Indeed, [2] shows how among 32 types of frequent risks, the highest probability of occurrence is related to quay crane accidents. Besides, [3] demonstrates how accidents occurring on quay cranes depends mainly on human error due to fatigue [3]. In this context, it is important to further explore this topic and to investigate if and how a fair management of quay crane operators may lead to better performances. The aim of this study was the simulation of the crane operator's fatigue. In particular, the Yerkes-Dodson law was embedded in a discrete-event multi-agent simulation environment. Specific scenarios were built and compared. Despite all the improvements and even the automation of several or all the elements in a port's sequence, human operators still play a central role on it [4].

Specifically, safety is considered as composed by two-parts, a passive part and an active one, each one regarding different aspects: active safety measures are activated to prevent an accident, whereas in the event of a collision, passive safety features are intended to soften the blow. Human factors can be classified as mechanical and psychophysical, each one being treated in the passive part and the active part, respectively. In order to reduce the negative effects of the human factors on performance, active safety focuses in how the perception of external stimuli is implemented, and how this information is managed by the mental processes underlying perception and finally the human decision making processes. Human factors that can be related with performance impair can be, for example:

- improper monitoring/lookout,
- drugs/alcohol consumption,
- fatigue,
- judgmental errors,
- negligence,
- regulations compliance flaws, etc.

Fatigue is thought to be a significant factor in accidents occurring in transport systems [5]. Many studies have been carried out to relate fatigue to operation performance. Furthermore, these studies can link work environment and conditions not only with performance impairment but also with a potential health hazard, both physical and mental. From a medical point of view, fatigue is distinguished from stress, as fatigue impairs performance, nevertheless stress is necessary up to a certain threshold for a person to be able to perform an activity well. In [6] the main types of stress factors are distinguished into physical and mental: physical stress factors are external or environmental stress factors such as heat, cold, vibrations; and mental ones consist of cognitive stress factors that depend purely on emotional factors of the person. The authors then present curves of human behaviour under the effects of fatigue and stress.

In the same way [4] studied the performance curves of crane operators in shift work. In this study, fatigue was the focus of the discussion. It was stated that fatigue, among other features, has physiological roots and was influenced primarily by sleep and the sleep-wake cycle, the circadian rhythm. The curves constructed with the average values of movements per hour corresponding to each shift could, in fact, be related to parabolic trends, being identifiable on different branches corresponding to hard tasks and easy tasks. Generally, the upper curve refers to an easy task and represents the high performance in the task development, while the lower curve concerns to difficult task and refers to low performance in the task development. These curves correspond to the Yerkes-Dodson law, it can be implied that the arousal stimulus is related with the time that the operators' amount of work done.

In the study proposed by [7] the authors focused on job stress and emotional impact on working conditions. A more social approach, it was attempted to spot the consequences, in terms of health and performance, of the work requirements assigned to the crane machinists. The experiment showed that almost 44% of the workers suffered of job strain, 49% were depressed and 90% suffered some level of anxiety, from mild to extremely severe. On the other hand, just 30% did have stress. Among the psychological factors, only anxiety was found significantly associated with job strain. Considering the independent factors analysed, an interesting result was obtained: physical isometric load and muscle ache were significantly associated with job strain.

The experimental study carried out by, [8], was performed by drivers and not crane operators. It is reportedly common among professional drivers or people in general, like tourists, to engage in considerably long driving activities after a deficient session of rest. This undoubtedly represents a major concern for traffic safety, as the traffic accidents records show ([9, 10]). It is important, then, to study the human limits regarding sleep deprivation and its effects on driving performance. Sleepiness is the desire of people to fall asleep, causing a strong difficulty to remain awake or conscious, and is triggered by poor sleep or by remaining awake for too long. It resolves after sleep. According to the results, there was an enormous difference between the two conditions. In fact, the number of errors for sleep-restricted (535) were 8 times bigger than the number of errors for the rested condition (66). A clear difference between the fatigue perceived in the two conditions was spotted: at the beginning of the session, the sleep-restricted subjects perceived almost twice the fatigue perceived by the rested subjects. Nevertheless, the variations over time were rather similar.

The last study reported has been carried out by [11], which drivers' behaviour on long trips was conveniently analysed. It is well known, for example, that truckers appreciate dedicated rest areas to stop in, besides the existing regulations about allowed driving times [12]. Drivers normally perform different activities while taking a break. These activities are, however, reduced to two possible conditions: the driver either remains seated or changes his posture to an upward one, more specifically walking. When talking about comfort, in driving terms, the first thing that comes in mind is where the driver's body rests, as posture is one important factor. The quality of the standard seats has increased considerably, as it's not considered a luxury anymore, evolving into more ergonomic designs, from a more posture-friendly perspective. The last aspect considered in the experiments was the duration of the driving sessions. The objective of this analysis was to try to prove the existence of a connection among the performance impairment, arousal, workloads, postures, breaks, sleep habit, work conditions, etc., in order to expose that several factors play a role in the operators' performance, and moreover, the deterioration of the health conditions.

Every study covered a different phenomenon and concluded with a way of improving the situation, in order to increase the overall production and performance in the port's system and the workplace conditions, as it shouldn't be acceptable to implement a work environment that undermines the workers' health. As the studies reflect, it should be a common exercise to check the potential hazards present in a work environment and put an effort into implementing appropriate improvements and protection. The objective of this study is to identify the curve representing the mental fatigue of quay crane operators in a real port.

As mentioned above, a lack of scientific contributions regarding the fatigue analysis of quay crane operators is observed in the available literature, to fill in this gap, in this study, a methodology based on the Yerkes-Dodson law (Y-D; [13]) in order to build the fatigue curve of crane operators is proposed. Additionally, this relation is implemented within a multi-agent discrete event and multi-agent simulation model of the Salerno container terminal. The framework of simulation model is object of a different study developed by the Transportation System Analysis group of the University of Salerno, this work is currently under review. The data considered for the analysis were provided by Salerno Container Terminal. The workload scenario was compared on the basis of appropriate performance indicators with respect to the do-nothing scenario. The results obtained showed the impact of fatigue through the performance curve. The paper is organized as follows: In the next section, the methodology applied to obtain the fatigue curves is presented, besides the specification of the quay cranes within the simulation model is described. Then the case study is presented as well as the data collection and the constructions of the fatigue curves for the simulation scenario. Further the results obtained for one shift are showed in the section four. Conclusions and the future prospects of the work end the paper.

2 Methodology

The methodology adopted is based on the Yerkes-Dodson law [14] which states that the optimal motivation for task performance decreases with increasing task difficulty (Fig. 1), also defined in literature as the inverted U model of arousal (Fig. 2). It states that at low arousal, performance is poor; it improves up to its highest point corresponding to the optimal level of arousal.

The left side of the curve represents low emotional arousal, the right side represents high emotional arousal and in the middle is a medium level of it. The vertical line on the left side goes from low performance (at the bottom) to maximum performance (at the top). The optimal state of emotional arousal and performance is located in the centre of the curve.

As shown in Fig. 1, the curve that refers to a simple task is the highest one, which represents a high performance, and the peak of the curve is further to the right than the curve that represents a complex task. Mathematically, these curves are represented by a second degree polynomial equation in which the coefficients change according to the degree of performance (container handling time) of the crane operators; thus when crane operators improve their skills to perform the tasks through training or gaining experience, this curve moves to the right (Fig. 3).

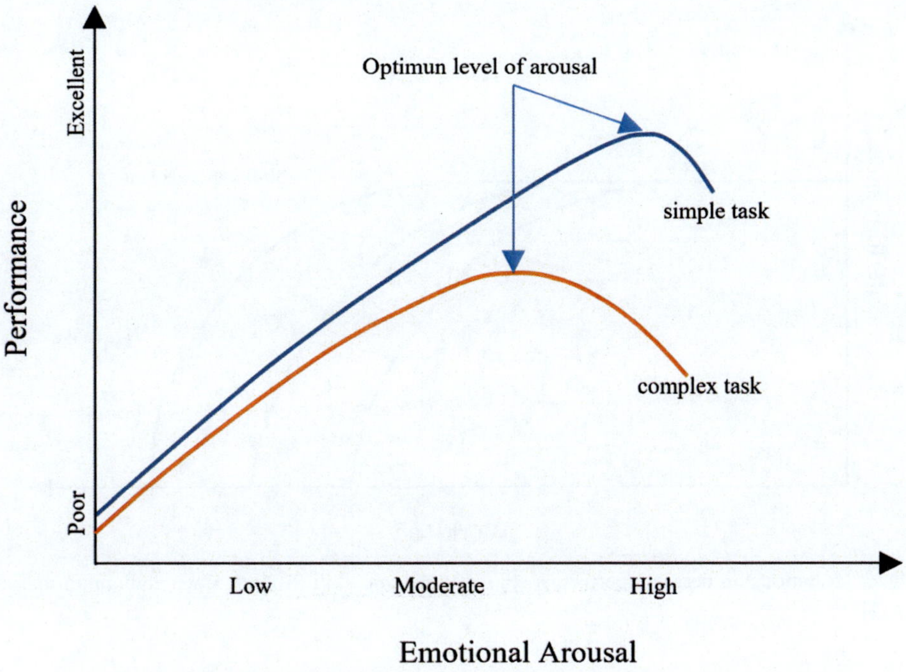

Fig. 1. Original Yerkes-Dodson law (Source: Own elaboration based on [14])

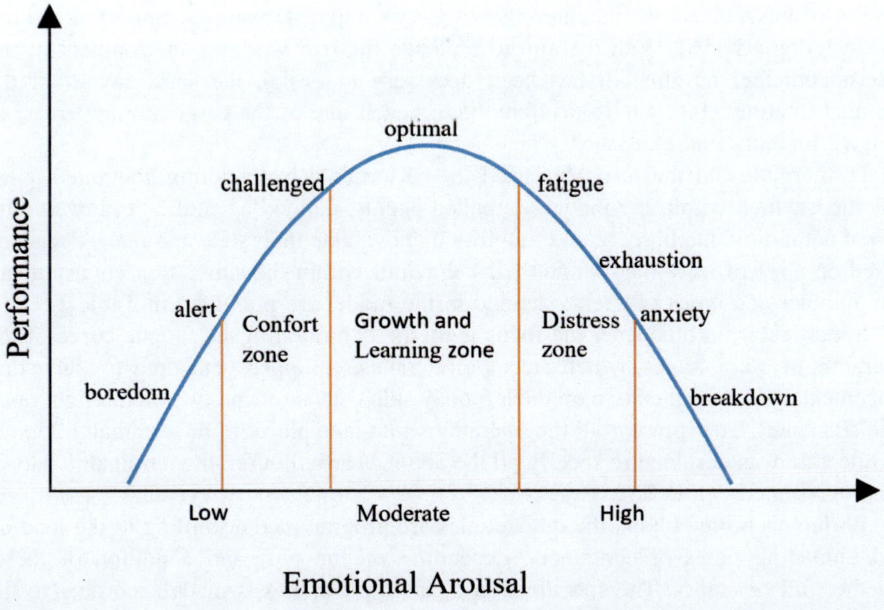

Fig. 2. Yerkes Dodson curve: Relationship between performance task and arousal/stress state (Source: Own elaboration based on [15])

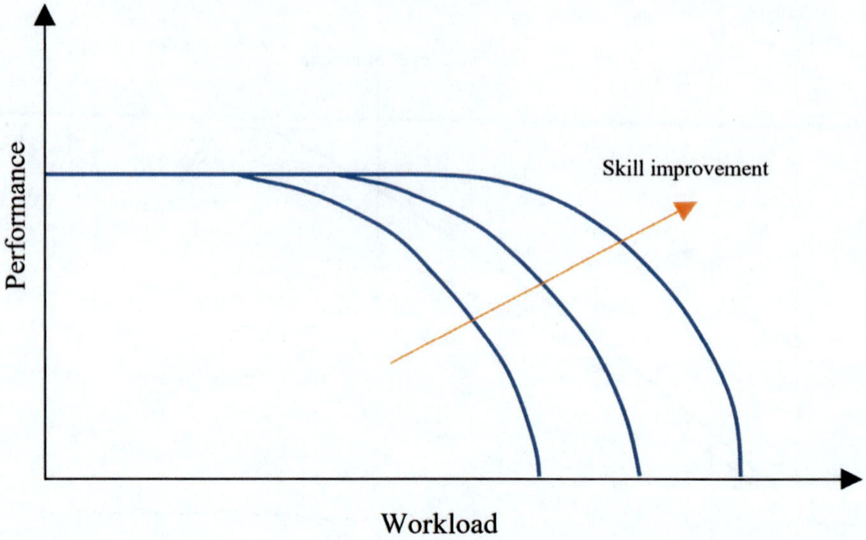

Fig. 3. Relationship between performance, workload and skill (Source: Own elaboration based on [4])

2.1 Model Simulation: Overview

The fatigue curve of the crane operators was implemented in a discrete event simulation model based on a multi-agent approach, in which the entire logistic process of freight loading/unloading and entering/exiting is simulated, using as support modelling tool AnyLogic® [16]. With the aim to replicate the real working environment of the Salernocontainer terminal, it has been necessary to design the same layout of the terminal in study (Fig. 4a, 4b), differentiating each one of the cargo storage areas, as well as the entry and exit lanes.

Furthermore, all the resources used in the terminal by handling containers were specified, which within the model are called agents, defined as entities endowed with partial autonomy, intelligence and mobility that evaluate their state and make decisions based on a set of rules that define their behaviour within the simulation environment. The number and types of agents defined in the model are presented in Table 1.

In particular, in this paper the focus is on the definition of the fatigue curve of the operators of quay cranes, which are mobile cranes, equipped with tired for their displacement and are located two on the left quay side (Levante) and two on the right quay side (Ponente). To represent all the operations that take place in the terminal, a logical architecture was designed to specify all the actions carried out in the terminal, a part of it is represented below Fig. 5.

Within each blue block, the quay cranes are programmed according to the loading and unloading times of containers, depending on the type and condition of these, whether full or empty. The specification of the quay cranes is made according to the fatigue curve of the crane operators, which will be defined in the next section.

a). b).

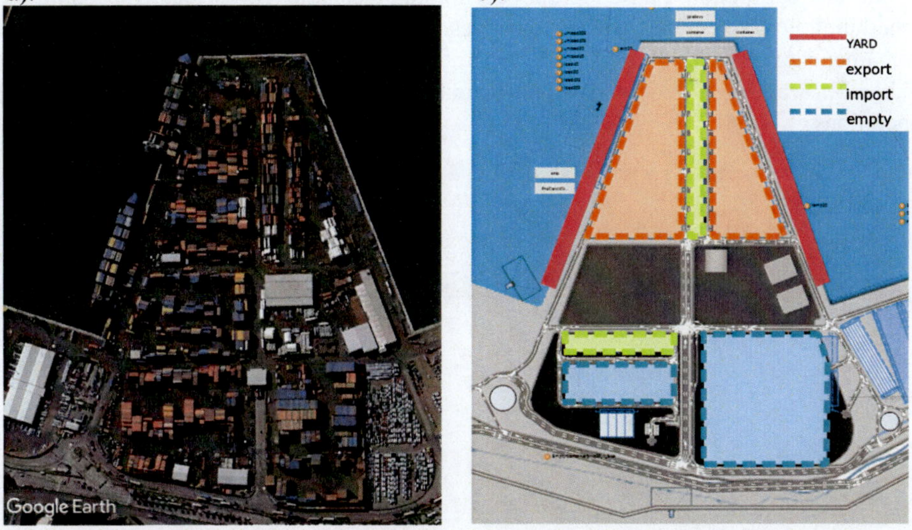

Fig. 4. (a). Layout of container terminal (Source: Own elaboration based on [17]), (b) its visual representation in the simulator

Table 1. Numbers and types of agents in the simulation model.

Agents	Quantity
Ships	One week scheduled
Quay cranes	4
Gantry cranes	2
Reach stackers	10
Tug masters	12
Trucks	One week scheduled

Table 2. Quay crane statistical results: parameters of Gamma distribution, undifferentiated container kind (Source: Own elaboration based on [18]).

Activity	Undifferentiated	
	μ	σ
Loading	1.366	0.514
Unloading	0.862	0.214
Loading from dock	1.389	0.441
Loading from shuttle	1.350	0.549
Unloading to dock	0.862	0.214

In the calibration phase of the do-nothing scenario (or scenario 0) and the scheduling both the cranes and the other resources with which the terminal carries out container movements, were used the operations time observed and reported in [18]. Tables 2, 3 and 4 shows quay crane handling times used, according to the type of operations.

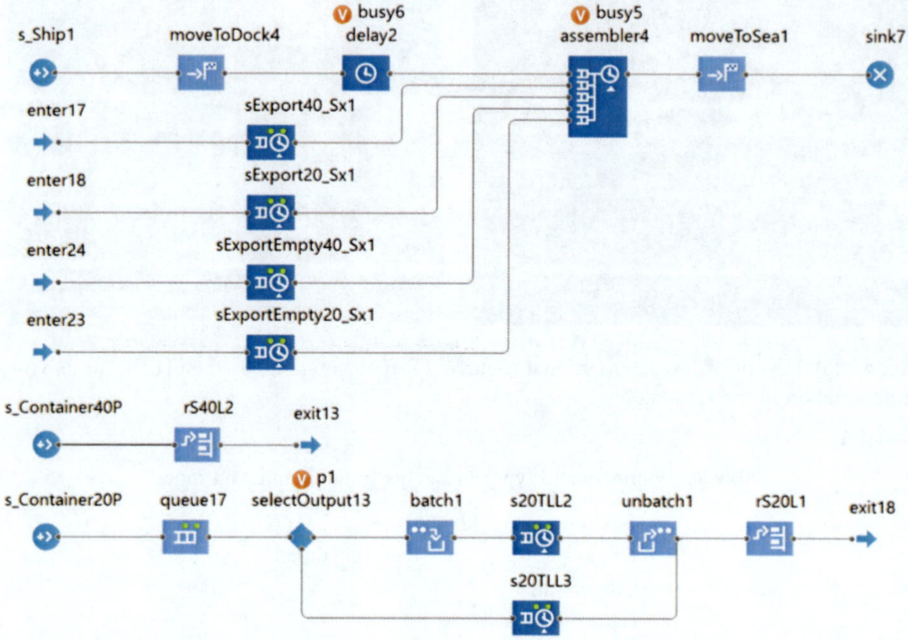

Fig. 5. Logical architecture of part/portion of the model

Table 3. Quay crane statistical results: parameters of Gamma distribution, 20 feet (Source: Own elaboration based on [18]).

Activity	20' Empty		20' Full	
	μ	σ	μ	σ
Loading	1.084	0.387	1.238	0.405
Unloading	0.664	0.139	0.825	0.183
Loading from dock	N.p	N.p	1.252	0.407
Loading from shuttle	1.084	0.387	1.227	0.405
Unloading to dock	0.664	1.139	0.825	0.183

Table 4. Quay crane statistical results: parameters of Gamma distribution, 40 feet (Source: [18]).

Activity	40'		40'	
	Empty		Full	
	μ	σ	μ	σ
Loading	1.101	0.340	1.288	0.402
Unloading	N.p	N.p	0.835	0.188
Loading from dock	N.p	N.p	1.372	0.485
Loading from shuttle	1.101	0.340	1.244	0.375
Unloading to dock	N.p	N.p	0.835	0.188

3 Case Study

3.1 Port of Salerno

The port of Salerno, located in the gulf of the Tyrrhenian Sea, is one of the largest national ports and one of the most efficient and dynamic ports in Europe. It is of great importance for the industrial and commercial system of central-southern Italy. From there it is possible to reach most places easily, thanks to the rapid connections with the highways. It ranks among the top regional ports for container handling. Ships access the port through a single inlet, of about 70 m wide. Regarding the commercial activity of incoming and outgoing containerised freight, this takes place in the "Trapezio" pier which is located in the centre of the port (Fig. 6).

Fig. 6. View of Salerno port (Source: Own elaboration based on [17])

The head berth can only accommodate medium-sized ships, while the others are long enough for the berthing of large ships. The existing harbour road system divides the pier into regular, wide, and deep storage yards. Those at the base of the pier do not have direct access to the quayside. Due to the characteristics described above, the "Trapezio" pier is therefore suitable for handling containers. These, in fact, require moorings suitable for large ships and close to each other handling and storage yards, in order to allow an effective and safe use of the mechanical equipment necessary for terminal operators in this field.

The container terminal works 365 days/year and 24 h/day in 4 shifts of 6 h each divided into:

- SHIFT: 08:00 ÷ 14:00,
- SHIFT: 14:00 ÷ 20:00,
- SHIFT: 20:00 ÷ 02:00,
- SHIFT: 02:00 ÷ 08:00.

The individual work shift is organised based on the available staff. Three types of operations can be identified, to which correspond as many different production cycles, with as many operational flows of freight and as many different input requirements:

- unloading of full container (import cycle),
- transhipment operations of full containers (transhipment cycle),
- loading of full containers (export cycle).

The limited space available in the yards has encouraged the research of solutions that could lead to the most intensive and efficient use of the space. The company's research resulted in two different solutions for the two main and different activities taking place in the storage areas:

- import cycle and empty delivery,
- export cycle and transhipments.

3.2 Data Analysis

The present research work is based on the data collection on movements both 20 and 40 feet containers either full or empty in the port of Salerno on six working days. These movements are clustered by work shifts, considering that each crane operator is subject to the activities of handling containers at full hours without considering partial working hours within each shift. The data collection is composed by moves by crane on the fullest work shift, that are the shift 2 and 3 (Table 5). The average of the movements per minute in each work shift was then calculated as well as the mean of the movements per hour of work in each shift. This will useful to generate an average fatigue curve for the operators in the work shifts.

Table 5. Data collection.

Shift	1	2	3	4	5	6	Mov/Min
2	0	11	17	16	4	0	
2	11	29	4	27	20	31	
2	31	44	41	36	26	17	
2	17	17	22	24	2	14	
2	20	23	26	13	22	5	
2	21	22	22	28	30	14	
Mean	20	24	22	24	17	14	0.34
3	7	23	18	21	13	11	
3	19	18	13	22	16	11	
3	7	7	10	11	7	8	
3	24	22	19	13	18	25	
3	11	17	31	27	18	8	
3	24	17	20	18	22	4	
Mean	15	17	19	19	16	11	0.28
Mean/hour	16	21	20	21	17	12	

3.3 Scenarios Simulation

Two scenarios are presented, scenario 0 or do-nothing scenario, which was specified, calibrated and validated by the Transportation System Analysis group of the University of Salerno and is therefore used as reference (real) scenario, and scenario 1 or workload scenario which is divided in 2 parts according to the approach followed in the way crane operators fatigue representation. The input data are reported in Table 6.

Table 6. Scenarios input data.

Day	Shift	Vessel time	Imp 40 F	Imp 20 F	Imp 20 E	Imp 40 E	Exp 40 F	Exp 20 F	Exp 20 E	Exp 40 E
04/06/2019	3rd	05:25	2	25	0	74	65	26	0	0

The operation times are represented by a Gamma distribution. From the study [18], these are chosen according to the activity carried out and the condition of the container (Tables 2, 3 and 4). The workload scenario was implemented through two approaches:

- the first one (1a) on deterministic base and,
- the second one (1b) through the application of the Yerkes-Dodson law.

In the first approach we attempt to implement a deterministic logic that simulates the fatigue curve of crane operators. There are various studies in the literature that build human performance curves as a function of time, the Gamma distribution implemented in the model [18], this takes into account time due to fatigue, however in this scenario we attempt to give to the fatigue a more central role. In AnyLogic® it was assumed as a deterministic value variable in the time. Assuming that the shift time is 6 h, it was decided to divide the shift into 3 parts: (i) the first, two hours of maximum efficiency where the operating time is given by the average (μ) minus the standard deviation (σ), (ii) the two central hours of medium attention where the processing time is equal to the average(μ), and (iii) the last two hours with a high level of fatigue where the processing time is given by the average(μ) plus the standard deviation (σ) (Fig. 7). Clearly, this hypothesis is a considerable simplification since a human task does not follow a deterministic time, however having reduced the time interval from 6 to 2 h it is more likely that the values assumed are close to the average.

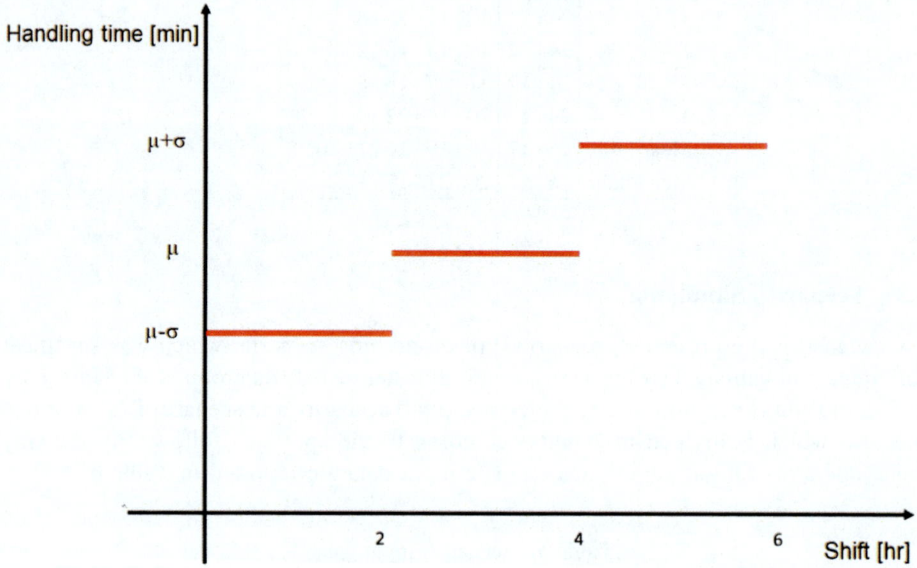

Fig. 7. Performance curve as function of time

In order to implement this logic, a state chart to three-stages has been constructed (Fig. 8). Cyclically, every 120 min, it passes from one stage to the next, modifying the processing times (Table 7). Due to the fact that the ships usually arrive at the beginning of the shift, this strategy takes advantage of the best moment of the workers by unloading most of the containers in the shortest time possible, then when the operator gets tired there are only a few containers left to process.

In the second approach, the performance was implemented according to the Yerkes-Dodson curves. Different curves are expected for each shift, as the stimulation differs between shifts, especially at night. As can be seen in (Fig. 9), the average performance curve for different shifts and different days constructed from data provided by Salerno Container Terminal has a maximum point almost halfway of the shift, with an increasing part to the left and a decreasing part to the right that is more pronounced than the previous one, indicating a strong influence of fatigue. The hypothesis described in deterministic scenario (1a) is not analogous to the idea presented in the theory of fatigue, which predicts maximum performance not at the beginning, instead after receiving the precise stimulation. Upon receiving more stimulation, in this case, performance tends to decrease more rapidly than it had increased.

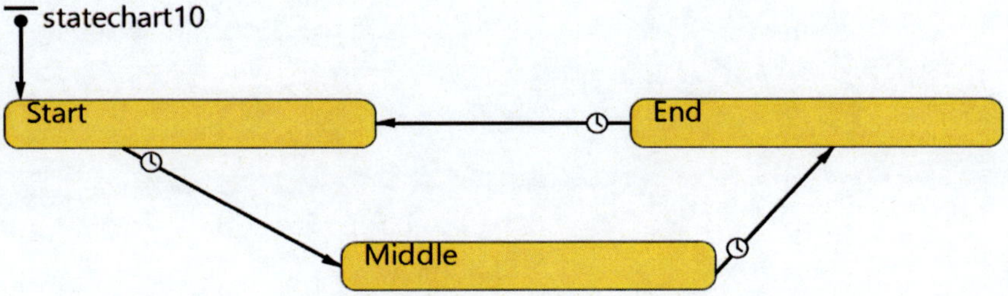

Fig. 8. Three stages state-chart

Table 7. Three stages processing times.

Activity	Start [min]	Middle [min]	End [min]
Unloading_Crane_Time_20_Full	0.642	0.825	1.008
Unloading_Crane_Time_40_Full	0.647	0.835	1.023
Unloading_Crane_Time_20_Empty	0.525	0.664	0.803
Unloading_Crane_Time_40_Empty	0.588	0.788	0.988
Loading_Crane_Time_20_Full	0.833	1.238	1.643
Loading_Crane_Time_40_Full	0.886	1.288	1.690
Loading_GC_Time_20_Full	0.430	0.741	1.052
Loading_GC_Time_40_Full	0.312	0.769	1.226
Unloading_RS_Time_20	0.088	0.144	0.200
Unloading_RS_Time_40	0.113	0.200	2.870
Loading_RS_Time_20	0.149	0.304	0.459
Loading_RS_Time_40	0.123	0.311	0.499
Stacking_RS_Time	0.114	0.260	0.406

The curve is represented by a second-order polynomial, and defined through Eq. (1).

$$Y = -1.107x^2 + 6.8857x + 10.567 \quad (1)$$

The coefficient of determination (R^2) of the (1) is equal to 0.94, and therefore resulted a good reliability in reproducing the phenomena. In this case, this behaviour is implemented within the software in a similar way to scenario (1a), but using data representing the trend described above. Corresponding values are then given to proportion the times that the quay cranes takes per hour; these proportion values are higher the more fatigue is felt (Table 8).

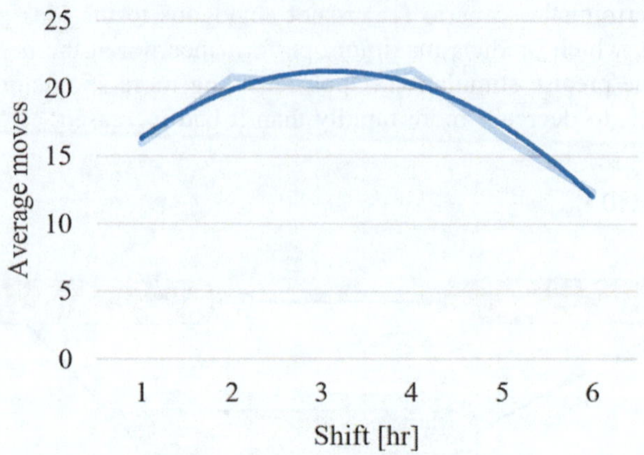

Fig. 9. Y-D curve created from the data of the average number of movements per shift

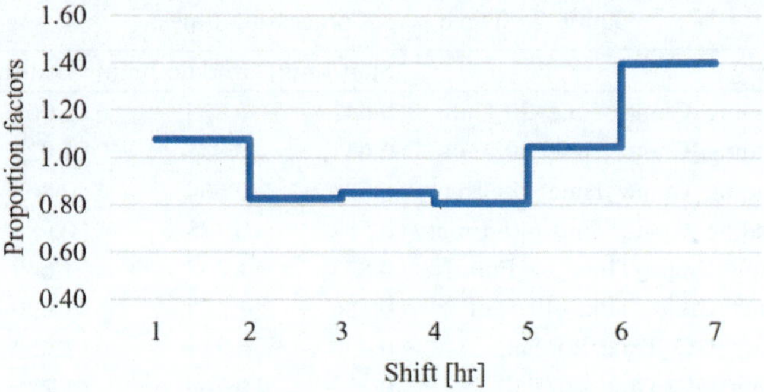

Fig. 10. Quay cranes working time proportion factors

Table 8. Quay Crane working times per hour. The times per hour are equal to the average for the proportion factors (Fig. 10).

Agents	Operation	Average(min)	Shift hours					
			1	2	3	4	5	6
			1.076	0.827	0.850	0.807	1.044	1.3960
Quay Crane	Load20E	1.084	1.167	0.896	0.922	0.807	1.131	1.513
	Load40E	1.101	1.185	0.910	0.936	0.889	1.149	1.537
	Unload20E	0.664	0.715	0.549	0.565	0.536	0.693	0.927
	Unload40E	0.835	0.899	0.690	0.710	0.674	0.871	1.166
	Load20	1.238	1.332	1.023	1.053	0.999	1.292	1.728
	Load40	1.288	1.386	1.065	1.095	1.040	1.344	1.798
	Unload20	0.825	0.888	0.682	0.702	0.666	0.861	1.152
	Unload40	0.835	0.899	0.690	0.710	0.674	0.871	1.166

4 Results

This section discusses the results of the simulated scenario. Scenario 0 is the starting point for this analysis. The ship time measured in reality is 325 min (data provided on the port of Salerno), while the average of the replications resulting from the statistical calculation is 314.31 min, with a difference of 3%, which is therefore a reliable, acceptable and representative value of reality. Regarding the scenario number (1a), the mean output is greater than the result measured in reality of 325 min, as expected, with a lower standard deviation than for scenario 0 (Table 9). Despite this, the error in this case is smaller, of 2%, it is closer to the real result, in a marginal proportion. The other statistical indicators are the same as before.

Table 9. Sample statistical analysis - scenario 0.

Run	Ship time(min.)
Run1	319.99
Run2	316.58
Run3	305.31
Run4	318.32
Run5	311.76
Run6	318.23
Run7	306.32
Run8	321.67
Run9	315.27
Run10	309.67

Number of initial replies	10
Average initial replies	314.31
std. dev. initial replies	5.74
confidence level	0.95
t Student value	2.262
Standard error	4.109
Absolute Precision	5
Minimum number of replies	7

Concerning the scenario 1a, it can be seen in the statistical results (Table 10), that it again comes out with a mean similar to the result measured in reality but still lower than that of scenario 0.

Table 10. Sample statistical analysis - scenario 1a.

Run	Ship time(min.)
Run1	331.51
Run2	331.26
Run3	332.94
Run4	329.75
Run5	328.49
Run6	333.80
Run7	332.81
Run8	333.39
Run9	337.35
Run10	332.08

Number of initial replies	10
Average initial replies	332.34
std. dev. initial replies	2.41
confidence level	0.95
t Student value	2.262
Standard error	1.726
Absolute Precision	2
Minimum number of replies	8

In this case (1b) (Table 11), the error is approximately 6%, practically twice that of scenario 0 and three times that of scenario (1a).

Table 11. Sample statistical analysis -scenario 1b.

Run	Ship time(min.)
Run1	302.38
Run2	301.12
Run3	305.11
Run4	302.28
Run5	304.78
Run6	305.67
Run7	303.83
Run8	305.57
Run9	299.49
Run10	310.22

Number of initial replies	10
Average initial replies	304.04
std. dev. initial replies	2.98
confidence level	0.95
t Student value	2.228
Standard error	2.004
Absolute Precision	3
Minimum number of replies	5

This should be related to the fact that when implementing this curve, the effect of fatigue is specifically taken into account, but not other events that have an influence on the total time of the ships, such as contingencies, failures, delays, etc. have been considered, whereas the Gamma distribution, implemented before, considered them. The poor performance of this configuration compared to scenario (1a) can be explained by the fact that, although configuration (1a) may seems over-simplified, the standard deviation includes the influence of other factors, making this modelling more accurate.

5 Conclusions and Future Prospects

The objective of this study was the simulation of the crane operator's fatigue. Several studies on workers fatigue were found in the literature but still is limited the research on crane operators. In this work, the Yerkes-Dodson law was used as reference for the representation of the fatigue behaviour for crane operators. The simulation model of the container terminal adopted in study is implemented using discrete event and multi-agent approaches on the AnyLogic software. A specific scenario was built to simulate operator's fatigue. This scenario was in two parts according the two following approaches: (1a) concerns a variation of the operating time within a work shift considering that the crane operator has his best performance at the beginning of the shift. The second approach (1b) covers a variation of the operating time following the Yerkes-Dockson law, which is represented by a convex polynomial curve, showing the operator's fatigue behaviour in a work shift.

In particular, the idea that the polynomial time-varying approach (1b) was a closer reproduction of reality than the deterministic time-varying approach (1a) was rejected, as the results showed that when implementing this curve, only the effect of fatigue was taken into account, but other events influencing the total ships operating time, such as accidents, breakdowns, delays, etc., have not been considered. On the other hand, although the configuration (1a) appears to be too simplified, nevertheless, by considering the time increasing through the standard deviation, the influence of other factors not taken into account by the other approach, is included. It is important to highlight that the investigation of the effect of fatigue must be analysed in depth, in order to cover the variables involved in the global definition of the problem, and thus to be able to apply this formulation accurately in the future. Finally, the model has always provided results that are considered encouraging considering the data provided on the port.

References

1. RMForwaring. http://rmforwarding.com
2. Andriani, D.P.: Risk analysis at ship to shore (sts) cranes in container terminal operational system of a greenport using failure mode and effect analysis. Int. Res. J. Eng. Technol. **4**, 2182–2187 (2017)
3. Fadda, P., et al.: Multidisciplinary study of biological parameters and fatigue evolution in quay crane operators. Procedia Manuf. **3**, 3301–3308 (2015)
4. Fancello, G., D'Errico, G.D., Fadda, P.: Processing and analysis of ship-to-shore gantry crane operator performance. J. Mar. Res. **5**, 39–58 (2008)
5. Fadda, P.: Biomeccanica e Antropometria. Introduzione al Fattore "uomo" Nella Fenomenologia dei Trasporti. University of L'Aquila, Italy (1984)
6. Seck, M., Frydman, C., Giambiasi, N., Ören, T.I., Yilmaz, L.: Use of a dynamic personality filter in discrete event simulation of human behavior under stress and fatigue. In: 1st International Conference on Augmented Cognition, pp. 22–27. Las Vegas (2005)
7. Yakub, N.W., Sidik, S.M.: Prevalence and contributing factors of job strain among crane operators in a port container terminal in Malaysia. Malaysian J. Med. Health Sci. **10**, 39–46 (2014)
8. Philip, P., et al.: Fatigue, sleep restriction and driving performance. Accid. Anal. Prev. **37**, 473–478 (2005)
9. Harrison, J.E., Mandryk, J.A., Frommer, M.S.: Work-related road fatalities in Australia, 1982-1984. Accident Anal. Prev. **25**: 453-451 (1993)
10. Personick, M., Mushinski, M.: Highway fatalities: leading cause of work-related deaths. Statistical bulletin (Metropolitan Life Insurance Company: 1984). **78**:19–25 (1997)
11. Sammonds, G.M., Mansfield, N.J., Fray, M.: Improving long term driving comfort by taking breaks - How break activity affects effectiveness. Appl. Ergon. **65**, 81–89 (2017)
12. Consiglio dell'Unione Europea: Regolamento (CE) n. 561/2006. https://eur-lex.europa.eu
13. Curry, D., Meyer, J., Jones, A.: Driver Distraction: Are We Mistaking a Symptom for the Problem?. SAE 2013 International Congress and Exposition (2013). https://www.sae.org/publications/technical-papers/content/2013-01-0439/
14. Boe, E.E., Church, R.M.: Century Psychology Series. Punishment: Issues and Experiments. Appleton Century Crofts, New York (1968)
15. Bodeker, G., Pecorelli, S., Choy, L., Guerra, R., Kariippanon, K.: Well-Being and Mental Wellness. Oxford Research Encyclopedia of Global Public Health. Oxford University Press, Oxford (2021)
16. AnyLogic Simulation Software. https://www.anylogic.com/
17. Google Earth. https://earth.google.com
18. Cartenì, A., de Luca, S.: Tactical and strategic planning for a container terminal: Modelling issues within a discrete event simulation approach. Simul. Model. Pract. Theory **21**, 123–145 (2012)

Preempting Fire Engines at Traffic Signals in Brunswick, Germany, Using ITS-G5

Sebastian Naumann(✉) and Joachim Schade

Institut f. Automation und Kommunikation e.V., Magdeburg, Germany
{sebastian.naumann, joachim.schade}@ifak.eu

Abstract. In case of emergency, professional emergency services should arrive at the point of destination as fast as possible. In urban areas, usually heavy traffic load and construction sites complicate the dispatching and coordination of emergency vehicles. Modern communication and information technologies continuously open up new possibilities, to bring emergency services safer and faster to their destinations. The paper presents research work on shorten the emergency trips of fire engines in Brunswick, Germany. The improvements focus on preemption of fire engines at traffic signals using ETSI ITS-G5 communication as well as route search based on historic GPS logs of emergency trips and current traffic parameters.

Keywords: Emergency vehicle · Traffic signal · Preemption · V2X · Route search · Fire engine · ITS-G5 · Emergency trip

1 Introduction

In case of emergency, professional emergency services should arrive at the point of destination as fast as possible. Many accidents, sustained injuries and other medical emergencies could be less severe, with many lives saved and not lost, if the firefighters and emergency medical assistance sought arrived in a timely manner; in some instances just seconds earlier. In urban areas, usually heavy traffic load and construction sites complicate the dispatching and coordination of emergency vehicles and prevent them from being faster. Moreover, emergency vehicles are involved in accidents on emergency trips year by year (Fig. 1). In particular, intersections are hotspots of crashes with other cars. Therefore, emergency vehicles are forced to slow down at these route sections in order to decrease the risk for crashes. For car and truck drivers on the other hand, it is not always easy to recognize from where the emergency vehicle is coming and where it will go.

Navigation solutions currently available on the market determine routes according to various target criteria (e.g. fastest or shortest route) with the help of data mining approaches. The route determination is based on the assumption that network sections have basically similar travel times on certain days of the week at certain times of the day. The existing approaches recognize, for example, accumulations of traffic disruptions, sections with regularly low traffic loads as well as much more complex relationships through the analysis of large amounts of historical traffic data. However, these algorithms fail in the event of unpredictable events, which in the short term lead to a

E. Macioszek and G. Sierpiński (Eds.): TSTP 2021, LNITI, pp. 120–132, 2022.
https://doi.org/10.1007/978-3-030-93370-8_9

strong deviation in the available transport (e.g. road closures, operational disruptions in local public transport) and/or in traffic demand (e.g. evacuation, diversion traffic, rescue logistics for large-scale operations). As a result, current navigation solutions are unable to conclude to the near future and thus about optimal routing decisions in these situations. In addition, such commercially available systems also lack the consideration of extended rights for special-purpose vehicles, such as driving through pedestrian zones or driving on one-way streets in the opposite direction. Local emergency services know the usual routes in their city and therefore usually the fastest route, something that many professional fire departments rely on. However, navigation systems are very practical for emergency services from neighboring regions, who are usually called in for support during major events. In addition, navigation solutions also provide support for emergency services with local knowledge if the general conditions change as a result of road closures or increased traffic.

Fig. 1. Crash of an emergency vehicle during an emergency trip (Professional fire brigade Munich, Germany)

Modern communication and information technologies continuously open up new possibilities, to bring emergency services safer and faster to their destinations. This paper presents research work on accelerating the emergency trips of fire engines in Brunswick, Germany. The improvements focus on preemption of fire engines at traffic signals using ETSI ITS-G5 communication and route search.

2 Related Work

Countless papers have been published around the world on guiding emergency vehicles safely and fast to the location of emergency. Particularly route search and preemption at traffic signals are major levers. Therefore, we only list some selected papers here. A comprehensive review of traffic signal preemption and priority for emergency vehicles from the perspective of the year 2001 with focus on technology provide Collura et al. [1].

Although preemption of emergency vehicles at traffic signals is applied all over the world, there are regional differences. Anil et al. [2] deal with the situation in India. Their study proposes a multilayer fuzzy model to determine the degree-of-priority (DOP) based on emergency vehicle preemption demand and impact intensity on each road section. Bycraft [3] proposes preemption at traffic signals for emergency vehicles in British Columbia, using different recognition technologies such as siren, strobe light and radio signals for transmitting and receiving the preemption requests.

With certain aspects of route search for emergency vehicles deal several research groups. Mohd et al. [4] introduces an interface of an A* algorithm using C# programming in order to find the shortest path for ambulances. Westergate et al. [5] analyzed travel time distributions of ambulances in large networks using a Bayesian method. The results are an important input for routing decisions. Also with travel time estimation deal Wang et al. [6], however, they focus on the influence of preemption control. They developed a model with three levels of preemption control strategies, including path preemption, intersection preemption, and section preemption.

3 Initial Situation

3.1 Protection Levels

In Germany, municipalities define so called protection levels for their fire brigades that minimally have to be reached [7]. Protection levels describe the quality of the danger prevention by the fire brigade in the community. The setting of the protection level is a political decision of the council. The levels of the fire protection plan should be set according to the Working Group of Heads of Professional Fire Brigades (AGBF). AGBF is a voluntary association of heads of professional fire departments in Germany. It is a self-supporting association in the German Association of Cities (DST).

When setting the protection level, quality criteria differentiated according to types of use are defined. In case of the fire brigades, each protection level describes:

- in what time (assistance period),
- with how much team and equipment (functional strength),
- in what percentage of cases (degree of target achievement),

they have to arrive at the location of danger. Furthermore, the fire protection plan includes:

- the description of the current state of the fire brigades in an administrative unit,

- the desired protection levels,
- a calculation of the degree of target achievement from the actual data,
- the need for action derived from the target/actual deviations in order to optimize the risk management capabilities to the extent agreed in the protection level.

In Brunswick, Germany, investigations revealed that above all emergency trips from the fire brigade main station to the western city did not arrive on point in too many cases. As a possible measure, the acceleration of the fire engines through the preemption of traffic signals was considered.

3.2 Speed Losses at Intersections

For many years, the professional fire brigade in Brunswick, logs the GPS datasets of their emergency trips. A key point on the way from the fire brigade main station to the western city is the square Rudolfplatz (Fig. 2). Since the fire brigade main station is in the north of Brunswick, a common scenario of getting to the western city is to pass the Rudolfplatz from the north to the south.

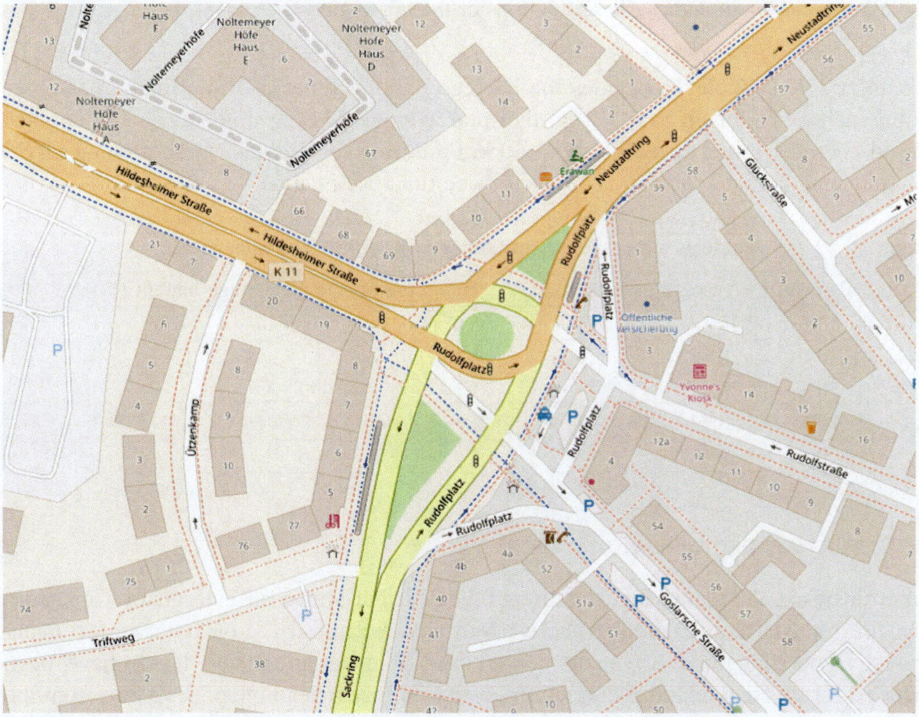

Fig. 2. Rudolfplatz (Rudolf square), Brunswick, Germany (OpenStreetMap contributors)

Figure 3 shows the average speed profile at the square Rudolfplatz during the recorded emergency trips. The speed drop in the middle of the square by more than

15 km/h can be clearly seen. There is no preemption of the fire engines at the traffic signal so far. The main reason for the speed reduction is therefore safety. Keeping the initial speed of about 60 km/h and more when entering the square would significantly increase the danger for crashes.

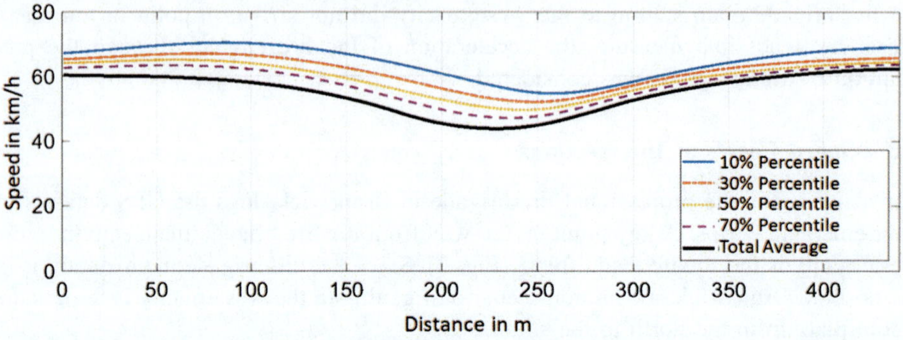

Fig. 3. Average speed profile at the intersection Rudolfplatz (Rudolf square) in Brunswick from the north to the south during emergency trips of the fire engines

In addition, Table 1 provides the average passing times at Rudolfplatz that differ from each other for up to 6 s from the total average passing time. If the fire engines could keep the initial speed of about 60 km/h during the passing without being afraid of accidents would probably save even more than 6 s of time.

Table 1. Average passing times at Rudolfplatz, Brunswick.

Percentile	Average passing time (s)	Time benefit (s)	Percentage benefit (%)
10%	26.76	5.67	17
30%	28.08	4.35	13
50%	29.12	3.14	9
70%	30.13	2.3	7
100%	32.43	–	–

3.3 Goals of Improvement

In Brunswick, Germany, investigations revealed that above all emergency trips from the fire brigade main station to the western city did not arrive on point in too many cases. Depending on the type of danger, two relevant protection levels have been defined for this trip relation. Protection level 1 defines a deadline of 9:30 min from the emergency call until the fire brigade arrives. Protection level 1 defines a deadline of 14:30 min from the emergency call until the arrival of the fire brigade.

In order to reduce the travel times of the fire engines on their emergency trips, (a) a preemption at traffic signals as well as (b) finding the fastest route was implemented.

The fastest route not only includes distance and speed of the road network but traffic parameters and roadworks as well.

4 Preemption at Traffic Signals

4.1 Beacons

In Germany, single vehicles have been preempted at traffic signals since the early 1970s by announcing them at request points. At such a request point, the vehicle sends out an infrared message to a radio beacon installed roadside. The traffic signal then switches to green after the predefined time a requesting vehicle needs to get from the notification point to the traffic signal. Unfortunately, all vehicle routes and switching times have been defined in advance. No dynamic information can be transmitted and processed. All actions are bound to the unique request point identifier. By means of the mostly rule-based control logic, the traffic light control is adapted to the expected arrival of the vehicle on the route specified in advance and supplied in the control. Such preemption systems are intended primarily to public transport vehicles. However, the principle is also occasionally used for emergency vehicles.

4.2 Traffic Control Center and GNSS

Siemens offers a preemption system called "Sitraffic Stream" [8], which uses receivers for satellite navigation (GNSS) in on-board units that are installed in special-purpose vehicles. The on-board units are connected to traffic control center via the mobile network. When a vehicle reaches a predefined position, a virtual notification point is triggered and a message is sent to the traffic control center. The traffic control then switches to a predefined emergency mode, turns the corresponding signal to green and only switches back to normal operation when the vehicle has left the intersection and the traffic control center has completed the preemption action. Conditionally compatible traffic flows are only taken into account if they have been supplied for emergency mode.

4.3 ETSI ITS-G5

ETSI ITS-G5 [9] describes the architecture and network mechanisms for V2V (vehicle to vehicle) and V2I (vehicle to infrastructure such as roadside units) communication as depicted in Fig. 4. The ITS-G5 architecture is based on some specific data types, such as the Cooperative Awareness Message (CAM) or the Decentralized Environmental Notification Message (DENM). These messages are generic and should be sent by all equipped vehicles without specifically addressing special-purpose vehicles. For special vehicles, however, additional parameters are possible, for example the notification of vehicles on an emergency trip with special signals.

Recently introduced have been the Signal Request Message (SRM) to be sent out by the vehicle and the Signal Status Message (SSM) [10, 11] to be sent out by the traffic signal, which both are dedicated to traffic signal preemption of public transport vehicles and emergency vehicles as well. Particularly compared to the radio beacons, the communication via ETSI ITS-G5 is much more flexible and, particularly, can inform connected passenger cars and trucks directly as well on the emergency trip taking place.

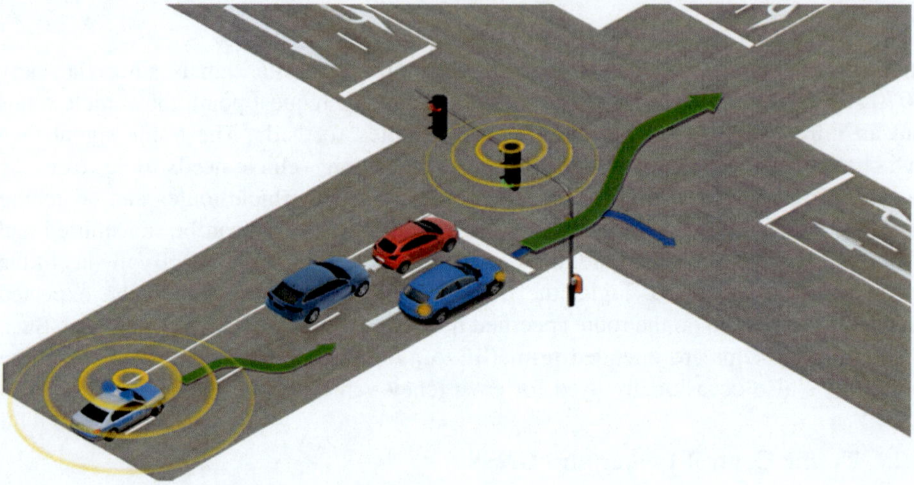

Fig. 4. Traffic signal preemption using ETSI ITS-G5 (Car2Car Communication Consortium)

All actions taken within this paper are based on the direct communication between the fire engines and the traffic signals using ETSI ITS-G5. This means, when a fire engine sends out a SRM the traffic signal control turns into a predefined emergency mode. Section 6 provides more details of the overall system and discusses some implementation issues.

5 Route Search

Beside traffic light preemption, finding the fastest route is the second block on emergency trip time reduction.

5.1 Database

The basic data for route search is stored in a PostgreSQL database. One table takes all nodes and one table takes all modified edges from the Open Street Map (OSM) road network. Modified means, compared to the original data taken from any OSM server where edges usually represent both directions, the edges are split such that one edge represents the forward direction and the second one the backwards direction (see Fig. 5).

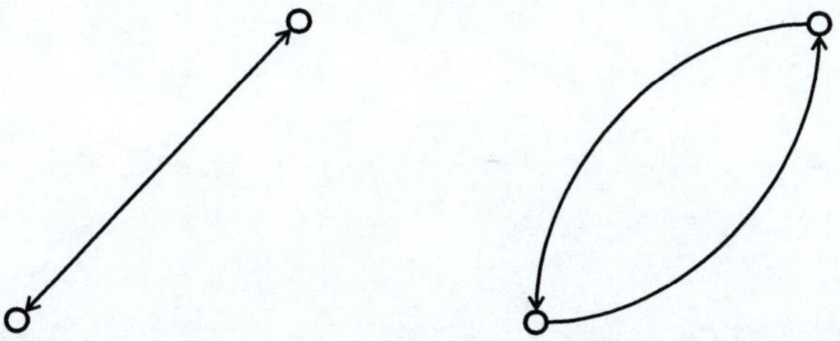

Fig. 5. Original OSM nodes and edge (left); original OSM nodes and split edge (right)

This structural modification has been done in order to have unique road network elements for traffic parameters like traffic density and traffic speed. Without this modification, each traffic parameter should have an additional flag indicating the direction. With respect to the route search algorithm, after this modification each edge can be traveled in one direction only. Now in the database tables we have one row for each node and one row for each edge (whereas there are almost twice as many edges compared to the number of original edges). Each edge dataset includes the geometry, such that curves can described, as well as the length in m and the speed limit.

Additionally, there a database tables containing the current and historic traffic parameters related to the edges: traffic speed, traffic density, historic speed profiles of fire engines. Finally, there is a table of signal-controlled intersections with and without preemption capability.

All locations of nodes and geometries are in WGS-84 format providing latitude and longitude.

5.2 Route Search Algorithm

Since the road network is limited to the Brunswick region, it consists of 71.123 nodes and 160.972 edges only. Related to today's personal computer's power, this is not much. Therefore, we just use Dijkstra's algorithm from 1959 [12] without any complicated speed-up techniques. Dijkstra's algorithm bases on a network of nodes and edges. Each edge connect exactly two nodes. Each edge is assigned with a cost value. In our case, the cost value is the travel time calculated by the edge's distance and speed. The speed is taken from historic emergency trips stored in the according database table. If an edge represents a part of an intersection with a signal control and the preemption of the fire engine is possible, the speed on this edge is kept high on the level when entering the intersection or in other words, no speed reduction is done.

Dijkstra's algorithm divides the nodes into the three groups "marked", "in queue" and "unvisited". Starting at the start node, all nodes directly reachable by this node (i.e. an edge is connected with it), are being visited now, and the cost to the start node is stored. Hereby, systematically the nodes' states changes from "unvisited" via "in queue" to "marked" such that a spanning tree is created. For each node is stored, through which edge it was added to the spanning tree. When the destination node is reached, the route finally is determined by tracing. Dijkstra's algorithm delivers the optimal route with a complexity of $O(|V|^2)$ where V is the number of nodes.

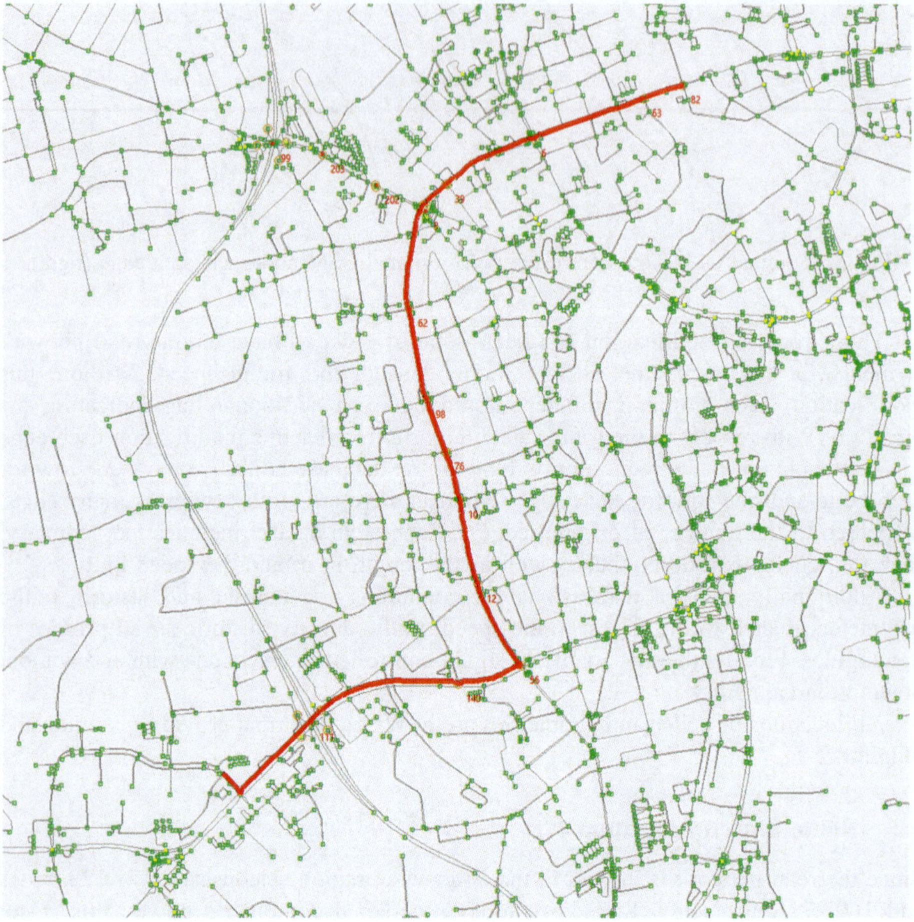

Fig. 6. Fastest route found

Figure 6 provides a route found within the test area. The red numbers represents intersections controlled by traffic lights and equipped with ITS-G5 technology.

5.3 Re-routing

Once, a fire engine has left its route by more than a predefined distance (e.g. 300 m), a re-routing from the current location of the fire engine to the destination is triggered. The first time route and the newly found route are provided to the fire engine via a MQTT broker. Details are described in the next section.

6 Implementation

6.1 System Architecture

The fire engines are equipped with an ITS-G5 communication unit (on-board unit). The traffic signal controllers are also equipped with an ITS-G5 communication unit (roadside unit). Based on these both unit the fire engines can directly send the signal preemption requests to the traffic signal controllers using standard internet protocols.

A central server runs the route search algorithm acting as a service. The fire engines are also connected to a central server via the mobile network. The fire engines report periodically its current locations to the central server. When an alert is triggered, route search is done for all fire engines involved from their respective locations to the emergency location. The routes found are sent back to the fire engines where they are displayed on a built-in monitor.

6.2 Data Distribution Using MQTT

MQTT is an OASIS standard messaging protocol for the Internet of Things (IoT). It is designed as an extremely lightweight publish/subscribe messaging transport that is ideal for connecting remote devices with a small code footprint and minimal network bandwidth. MQTT today is used in a wide variety of industries, such as automotive, manufacturing, telecommunications, oil and gas, etc. [13]. In our system, a MQTT broker has been set up on the central server. We use the open source Eclipse Mosquitto [14] on a Linux operating system. Publishers and subscribers are connected by topics. In this sense, the fire engines publish their locations on a certain topic (e.g. "sirene/vehicles"), and the routing service publishes the computed routes on another topic (e.g. "sirene/routes"). The routing service subscribes the topic "sirene/vehicles" whereas the fire engines subscribe the topic "sirene/routes". Figure 7 shows this principle clearly. The JSON data format is used for all messages transmitted via the MQTT broker. An example of a route message is provided in Fig. 8.

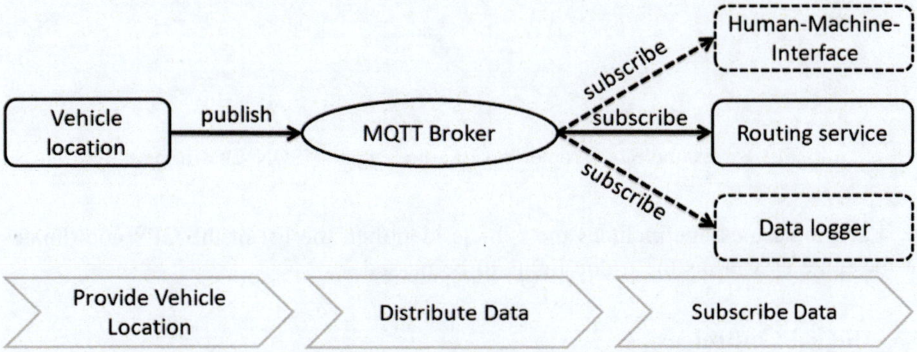

Fig. 7. Architecture of the MQTT data distribution model

```
{
  "type": "FeatureCollection",
  "features": [{
    "type": "Feature",
    "properties": {
      "tourid": "1180103623",
      "resource": "BS 01-48-01",
      "ts": "1550766543",
      "eta": "1550767108"
    },
    "geometry": {
      "type": "LineString",
      "coordinates": [
        [10.506909, 52.272453],
        [10.506910, 52.272454],
        [10.506915, 52.272460],
        [10.500918, 52.270069],
        [10.506346, 52.255186]
      ]
    }
  },
  {
    "type": "Feature",
    "properties": {
      "lsa_id": "lsa_10"
    },
    "geometry": {
      "type": "Point",
      "coordinates": [10.506909, 52.272453]
    }
  },
  ...
  ]
}
```


Fig. 8. Example of a route sent to a fire engine (JSON data format)

Each route message includes the vehicle identifier, the list of the GPS coordinates of the route as well as the traffic lights to be passed.

6.3 Signal Control

Despite the much more flexible signal request via ITS-G5, the signal control must be prepared on emergency vehicle preemption. This requires the definition of special signal programs resp. phases for all possible incoming lanes of an intersection. A very simple approach is to switch to red the signal groups of all incoming lanes without any exception. However, this may block the way for the emergency vehicle at high traffic density through waiting cars and trucks. Therefore, a better approach is to switch to green the signal group of the incoming lane the emergency vehicle is going to use, because possible blocking cars and trucks have the chance to clear the way.

6.4 Issues

Implementing, testing and evaluating systems like the one presented in this paper brings some difficulties. Traffic signal operation is a sovereign task of the responsible municipality even though all hard and software is ordered from and installed by private companies. Malfunctions can have dire consequences like car crashes with dead and injured people. Therefore, traffic signal operators are very carefully according to research activities. All changes on the signal control have to be critically reviewed within tedious processes. Unfortunately, open programming interfaces are rare or not available. On one hand, this is understandable from the operator's point of view, but on the other hand, a more open system architecture with a safety and security layer could simplify the research in this area with a meaningful social benefit.

7 Results and Conclusions

The system described within this paper has been implemented in Brunswick at selected signal controlled intersections. The evaluation has just started, and it will last until the end of 2021. At least, the test drives and the few emergency trips performed so far have shown a continuously higher speed compared to the ones without the developed system.

Not considered so far are concurrent signal preemption requests of rescue teams hurrying to different emergency locations at the same time. Of course, this issue occurs very seldom but should be a future research topic anyway.

The ITS-G5 communication technology is much more flexible than previous technologies. A major benefit of ITS-G5 is that connected cars and trucks (i.e. they own an ITS-G5 communication module) are informed directly about an approaching emergency vehicle. On-board assistance systems can support the drivers to behave correctly.

Acknowledgements. The work presented in this paper has been carried out within the German national research project SIRENE funded by German Federal Ministry of Transport and Digital Infrastructure under the grant number 19F2030A. The authors would like to thank for the financial support.

References

1. Collura, J., Chang, J., Willhaus, E.W., Gifford, J.: Traffic signal preemption and priority: technologies, past deployments, and system requirements. In: ITS America 11th Annual

Meeting and Exposition, ITS: Connecting the Americas Intelligent Transportation Society of America (ITS America) (2001). https://trid.trb.org/view/693552

2. Anil, R., Satyakumar, M., Jayakumar, J.: Travel time estimation and routing for emergency vehicles under Indian conditions. In: 3rd IEEE International Conference on Intelligent Transportation Engineering (ICITE), pp. 247–252 (2018). https://doi.org/10.1109/ICITE.2018.8492696
3. Bycraft, J.R.: Green light pre-emption of traffic signals for emergency vehicles Richmond, British Columbia's approach. City Richmond Traffic Signal Control Syst. (2013). https://www.richmond.ca/__shared/assets/Route_Preemption9281.pdf
4. Nordin, N.A.M., Zaharudin, Z.A., Maasar, M.A., Nordin, N.A.: Finding shortest path of the ambulance routing: interface of A* algorithm using C# programming. In: 2012 IEEE Symposium on Humanities, Science and Engineering Research, Lumpur, Malaysia. IEEE Press (2012). https://doi.org/10.1109/SHUSER.2012.6268841
5. Westergate, B.S., Woodard, D.B., Matteson, D.S., Henderson, S.G.: Large network travel time distribution estimation for ambulances. Eur. J. Oper. Res. **252**(1), 322–333 (2016)
6. Wang, J., Yun, M., Ma W.: Travel time estimation model for emergency vehicles under preemption control. In: 13th COTA International Conference of Transportation Professionals CICTP, vol. 96, pp. 2147–2158 (2013)
7. Protection level. https://de.wikipedia.org/wiki/Schutzziel_(Brandschutz). (in German)
8. Satellite-based preemption Sitraffic Stream, Siemens. https://assets.new.siemens.com/siemens/assets/api/uuid:3017d25b9274a90df129246ac062ff7f723e7674/sitraffic-stream-fire-brigade-v2-de-180223.pdf
9. ETSI ES 202 663: European Profile Standard for the Physical and Medium Access Control Layer of Intelligent Transport Systems Operating in the 5 GHz Frequency Band, ver. 1.1.0 (2009)
10. Dutch Profile for the Signal Request Message and Signal Status Message. CROW (2020). https://www.crow.nl/downloads/pdf/verkeer-en-vervoer/verkeersmanagement/landelijke-ivri-standaarden/d3046-3_srm-profile.aspx
11. V2X Communications Message Set Dictionary. https://www.sae.org/standards/content/j2735_202007/
12. Dijkstra, E.W.: A note on two problems in connexion with graphs. Numerische Mathematik **1**(1), 269–271 (1959)
13. Message Queuing Telemetry Transport (MQTT). https://mqtt.org/
14. Eclipse Mosquitto. https://mosquitto.org/

A Numerical Case Study on the Thermal Runaway of a Lithium-Ion EV Battery Module

Aleksander Król[1(✉)] and Małgorzata Król[2]

[1] Faculty of Transport and Aviation Engineering, Silesian University of Technology, Katowice, Poland
aleksander.krol@polsl.pl

[2] Faculty of Energy and Environmental Engineering, Silesian University of Technology, Gliwice, Poland
malgorzata.krol@polsl.pl

Abstract. Lithium-ion batteries are nowadays the main energy source for electric vehicles. Although the general principle of their operation remains unchanged for dozen of years, manufacturers try to achieve as much as possible energy density, hence many improvements have been introduced. On the other side, lithium-ion batteries are prone to failures, which may lead to a chain reaction in the whole battery module. Since the energetic outcome of such event can pose a serious threat to people, the data on the rate of the chain reaction development is important due to safety issues. Hence, the paper presents a detailed numerical study on thermal runaway in battery modules. Reactions in a single cell were modeled in accordance to up to date knowledge and then the propagation of the failure through a battery module was analyzed. The results showed the reaction might develop in a latent manner and culminate violently after a long time period.

Keywords: Lithium-ion batteries · CFD · Electric vehicle · Thermal runaway

1 Introduction

The world of motorization is changing now. After over one hundred years of domination, vehicles powered by internal combustion engines (ICE) begin to give way to electric vehicles (EV). It is because the economies based on fossil fuels are facing serious problems [1]. Therefore, mainly due to environmental protection issues, most of all the need of decrease of green house gases and other pollutants emission, the observed tendency will grow stronger. This should mitigate the climate change [2] and improve air quality, especially in cities [3]. However, the widespread use of EVs faces some obstacles, which are connected with the amount of electric energy to be stored to ensure a reasonable driving range. The most common solution is to apply lithium-ion batteries [4]. Despite a rapid development, the energy density obtained for lithium-ion batteries is still significantly lower than for hydrocarbon fuels. It is at most 1.4 MJ/kg [5], which is about 2% of the energy density of gasoline. This forces the big sizes of battery packs and causes slow recharging process [6]. Additionally batteries undergo

E. Macioszek and G. Sierpiński (Eds.): TSTP 2021, LNITI, pp. 133–146, 2022.
https://doi.org/10.1007/978-3-030-93370-8_10

some processes, which can lead to their age degradation [7]. This manifests by a lowered battery capacity after a number of charge and discharge cycles. The capacity of a lithium-ion battery decreases in low temperatures [8]. Moreover, such circumstances may even cause some irreversible unwanted changes. But, indisputably the most serious risk may arise if some violent exothermal process outbreak inside a battery cell [9]. This can be caused by an abuse of mechanical, electrical or thermal nature and may lead even to an explosion [10].

The aim of the presented work is to trace the process of thermal runaway in a car battery module and to estimate the time between the primary failure to the possible dangerous effects of the high energy release. The rest of the paper is arranged as follows: first the most important reactions, which may occur in a out-of-order battery cell are described. Then, the overall structure of analyzed numerical model is presented. In the Results section the outcomes of the carried out numerical experiments for a single cell and for entire battery module are described. Finally, the Conclusion section sums up the most important findings.

2 State of Art

Although there are many varieties, typical single lithium-ion cell contains three basic components: a graphite anode, a cathode made of a kind of lithium oxide and an electrolyte, which is a solution of a lithium salt in an organic solvent. Since the distance between the anode and cathode should be as small as possible to assure the battery efficiency, a thin plastic sheet is inserted to act as a separator [4]. The separator is perforated enabling ions to pass, while keeping the electrodes apart. Commonly, sheets of both electrodes and the separator soaked with the electrolyte are rolled up forming a jellyroll. Although, all electrochemical reactions in a lithium-ion cell are very complex, generally during the charging process lithium ions move from the cathode to the anode. Meanwhile, electrons forming the electric current flow through an external electric circuit. When a battery is discharged the process is reversed [11]. During the first charging/discharging cycle a layer between the electrodes and the electrolyte is formed. This layer is called Solid Electrolyte Interface (SEI) and plays an important role in a cell operation [12].

The normal operation of a lithium-ion battery is a balance among many component processes. Their overall efficiency is not ideal and an amount of energy is lost. The losses are of order of 1% [13]. The lost energy is transformed into heat. The amount of such generated heat should be equal to the amount of the dissipated heat, which is a condition of the stable battery operation. Under some circumstances this balance is disrupted and the temperature inside a cell may increase. If the temperature exceeds given thresholds, some undesired exothermic reactions may be triggered. These reactions undergo the Arrhenius law, which describes the dependence of a reaction rate (r_r)

on the absolute temperature (T). Since these reactions are regarded as of first order type, r_r expresses the number of molecule collisions that result in a reaction per second per volume unit:

$$r_r = a_r e^{-\frac{E_{ar}}{kT}} \tag{1}$$

In formula (1) E_{ar} [J/kg] denotes so called activation energy, which also depends on the reaction; k is Boltzmann constant. The term a_r [1/s] is called frequency factor and denotes the overall number of molecule collisions per second and volume unit. It may be regarded as a constant, which depends on the reaction. As it can be seen the rate of a reaction increases with the temperature. If a reaction is of exothermic nature, it may result in the temperature increase when cooling is insufficient. This positive feedback could quickly reach a 'no return point' and lead to a self-accelerated reaction.

Different sources mention many reactions, which may occur when 'no return point' is reached. The detailed literature study allowed for pointing out the most important of them [6, 14, 15]. Each of the presented processes is described by a couple of equations, first determining the reaction rate, second determining the rate of heat generation per volume unit ($\dot{Q}$). All of the used symbols are shown and explained in Table 1. The given values concern typical 18.650 battery (more specifically: Panasonic NCR 18.650-B).

When the temperature exceeds 70 °C the reaction of decomposition of solid electrolyte interface (SEI) begins.

$$\frac{dC_{SEI}}{dt} = -C_{SEI} a_{SEI} e^{-\frac{E_{SEI}}{kT}} \tag{2}$$

$$\dot{Q}_{SEI} = -h_{SEI} W_{SEI} \frac{dC_{SEI}}{dt} \tag{3}$$

Now, when the SEI has been decomposed, at the temperature of 120 °C a direct reaction between the anode and the electrolyte (negative-electrolyte, anode decomposition) may occur.

$$\frac{dC_{NE}}{dt} = -C_{NE} a_{NE} e^{-\frac{E_{NE}}{kT}} e^{-\frac{z}{z_0}} \tag{4}$$

$$\frac{dz}{dt} = C_{NE} a_{NE} e^{-\frac{E_{NE}}{kT}} e^{-\frac{z}{z_0}} \tag{5}$$

$$\dot{Q}_{NE} = -h_{NE} W_{NE} \frac{dC_{NE}}{dt} \tag{6}$$

Similarly, when the temperature exceeds 200 °C, a direct highly exothermic reaction between the cathode and the electrolyte (positive-electrolyte) starts.

$$\frac{d\alpha}{dt} = \alpha(1-\alpha)a_{PE}e^{-\frac{E_{PE}}{kT}} \tag{7}$$

$$\dot{Q}_{PE} = h_{PE}w_{PE}\frac{d\alpha}{dt} \tag{8}$$

If a cell is charged to some degree, it is possible to release the energy due to a short circuit. It may occur when the separator melts up (about 165 °C).

$$\frac{dSoC}{dt} = -SoCa_{SoC}e^{-\frac{E_{SoC}}{kT}} \tag{9}$$

Table 1. Parameters of the thermal runaway processes (Source: Own research based on: [6, 14, 15]).

Process/acronym	Symbol	Description	Value/initial value
SEI decomposition (SD)	C_{SEI}	Fraction of lithium in SEI	0.15
	a_{SEI}	Frequency factor	$1.667 \cdot 10^{15}$ [1/s]
	E_{SEI}	Activation energy	$2.24 \cdot 10^{-19}$ [J]
	h_{SEI}	Heat release	$2.57 \cdot 10^{5}$ [J/kg]
	w_{SEI}	Specific content of carbon in jellyroll	$6.104 \cdot 10^{2}$ [kg/m^3]
Reaction of negative-electrolyte (NE)	C_{NE}	Fraction of lithium in the anode	0.75
	a_{NE}	Frequency factor	$2.5 \cdot 10^{13}$ [1/s]
	E_{NE}	Activation energy	$2.24 \cdot 10^{-19}$ [J]
	z	Dimensionless SEI thickness	0.033
	h_{NE}	Heat release	$1.714 \cdot 10^{6}$ [J/kg]
	w_{NE}	Specific content of carbon in jellyroll	$6.104 \cdot 10^{2}$ [kg/m^3]
Reaction of positive-electrolyte (PE)	α	Degree of the cathode conversion	0.04
	a_{PE}	Frequency factor	$6.667 \cdot 10^{13}$ [1/s]
	E_{PE}	Activation energy	$2.03 \cdot 10^{-19}$ [J]
	h_{PE}	Heat release	$3.14 \cdot 10^{5}$ [J/kg]
	w_{PE}	Specific content of positive active in jellyroll	$1.438 \cdot 10^{3}$ [kg/m^3]

(*continued*)

Table 1. (*continued*)

Process/acronym	Symbol	Description	Value/initial value
Short circuit (SC)	SoC	State of charge	100%
	a_{SoC}	Frequency factor	$3.37 \cdot 10^{12}$ [1/s]
	E_{SoC}	Activation energy	$1.58 \cdot 10^{-19}$ [J]
	h_{SoC}	Heat release per cell	$10.17 \cdot 10^{3}$ [J]
	V_{cell}	Cell volume	$1.663 \cdot 10^{-5}$ [m^3]
Electrolyte decomposition (EC)	C_{EC}	Dimensionless concentration of electrolyte	1
	a_{EC}	Frequency factor	$5.14 \cdot 10^{25}$ [1/s]
	E_{EC}	Activation energy	$4.55 \cdot 10^{-19}$ [J]
	h_{EC}	Heat release	$1.55 \cdot 10^{5}$ [J/kg]
	w_{EC}	Specific content of electrolyte in jellyroll	$4.069 \cdot 10^{2}$ [kg/m^3]

$$\dot{Q}_{SoC} = -\frac{h_{SoC}}{V_{cell}} \frac{dSoC}{dt} \tag{10}$$

Finally, above 230 °C the reaction of electrolyte decomposition begins.

$$\frac{dC_{EC}}{dt} = -C_{EC} a_{EC} e^{-\frac{E_{EC}}{kT}} \tag{11}$$

$$\dot{Q}_{EC} = -h_{EC} w_{EC} \frac{dC_{SEI}}{dt} \tag{12}$$

The marking 18.650 means just the cell dimensions: it is a cylinder of 18 mm diameter and 650 mm height. Beside this, these batteries may differ in many details of inner structure, however the general principle of operation remains the same. Figure 1 shows all the processes taken into account and the temperatures which trigger them.

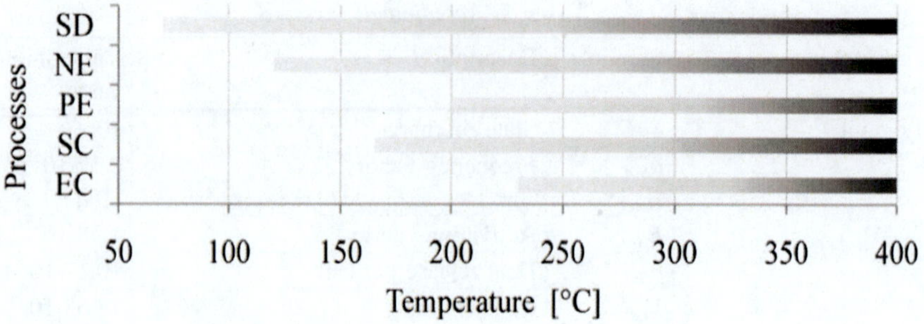

Fig. 1. The overall diagram of examined processes

If the heat is not dissipated from cell in a sufficient rate, even a small excess of the temperature will trigger the first of these processes, and then subsequent ones will be activated. This is called thermal runaway.

A car battery module contains a large number of single cells. The cells are packed together tightly in a chassis. The number of cells depends on manufacturer's detailed solutions and vehicle model. For instance a Tesla car battery module contains over such 7000 cells [14]. Many experimental and theoretical research showed that a failure of a single cell may cause a chain reaction and lead to an abnormal increase of the temperature of the whole battery module [16, 17]. It could manifest as a explosion and a violent fire. The initial failure may have different reasons - it could be a result of a mechanical punching, short circuit, overheating or general failure of battery management system. Irrespective of an original cause, the thermal runaway process goes in a similar way. The increased temperature in the primarily damaged cell triggers unwanted reactions in directly neighboring cells. This in turn, as their inner temperature increases may be the beginning of the chain reaction, which starts to propagate over the entire module.

Even a single battery cell is a complex system, so the propagation of the thermal abuse over the entire module is the more complicated phenomenon, which still requires detailed studies. Moreover, if there is an attempt to extinguish a fire of such battery module, the most important activity is to completely dissipated the generated heat. Otherwise, the chain reaction will run up again [18].

3 The Numerical Model

The numerical model was built using ANSYS Fluent. It is a well recognized software package, which numerically solves the tasks of fluid and heat flow. ANSYS Fluent implements Finite Volumes Method (FVM), in which the whole computational domain is divided into small volumes. For each such volume, balance equations are solved for all quantities describing fluid and heat flow (mass, energy, momentum). Since in the considered case, the electrolyte is a viscous fluid [19], which soaks the micro-perforated separator, just the laminar model of flow used.

The reactions, described by Eqs. (2)–(12) were introduced into the numerical model using User Defined Functions (UDF). An UDF is an ANSYS Fluent feature, which allow for expanding the software capabilities. It is a dedicated procedure written in C language and then plugged-in into relevant slots of the numerical model.

All exothermic reactions are implemented by a single UDF, which taking into account the actual temperature of each finite volume calculates the amount of released energy.

Different battery modules containing typical 18.650 cells were modeled. A single cylindrical cell is modeled as a homogeneous porous zone. The cell walls are made of thin steel sheet. The physical properties of a cell are shown in Table 2. The density, the thermal conductivity and the specific heat of the jellyroll, which fills a cell are the effective values: they are mass averaged properties of electrodes, electrolyte and separators [20].

Table 2. Properties of a single cell model.

Property	Value	Unit	Property	Value	Unit
Diameter	18	mm	Specific heat	830	J/kgK
Height	650	mm	Density	2580	kg/m^3
Capacity	3600	mAh	Thermal conductivity	3.4	W/mK

The chassis of the battery module is made of aluminum. There are two possible arrangements of cylindrical battery cells: aligned, in which cells are arranged in regular rows and staggered, in which neighboring rows are shifted. Both arrangements are similar in respect of space utilizations [21]. In the presented work the aligned arrangement was modeled, as is shown in Fig. 2. Additionally, cells markings used later in the article are shown as well.

The initial temperature of the whole assembly was 27 °C (300 K). The outer walls of the battery module was assumed to be cooled by a free stream of air of the temperature 27 °C too. The 'thermal convection' wall sub-model was applied here with the heat transfer coefficient equal to 35 W/m^{2}K.

Since the dynamic phenomena were studied, the simulation were naturally carried out in the transient mode. The time step was varied during simulations: initially it was set to 0.1 s, which in series of preliminary tests appeared small enough to accurately catch the time evolution of the SD, NE and SC processes.

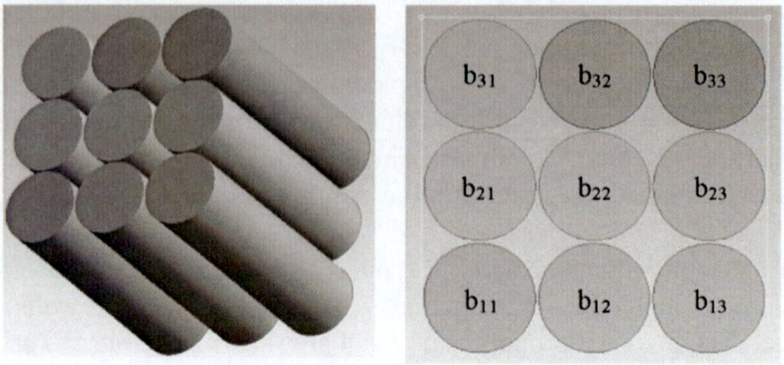

Fig. 2. The aligned arrangement of battery cells in the module model (left). The markings of the cells (right).

When the PE and EC processes started to develop, the time step had to be shortened even to 0.0001 due to their extra violent nature. Initially in every following numerical test, a failure in one of the cells was triggered and then the progress of the chain reaction was studied. The initial damage was simulated by heating up one of the cells. The temperature increase was high enough to start the first of the series of exothermic reactions - the decomposition of SEI (SD).

4 Results

At the first stage of the research the process in a single cell were examined, which was an experiment similar to real test carried out by Coman [20]. The heat flow provided to the cell was equal to 100 W. Figure 3 presents the calculated average cell temperature, which is compared with real data coming from Coman's experiment [20]. As it can be seen both temperature curves run in a similar way. There are some differences visible in the initial phase of battery warming up. They resulted from some simplifications introduced to the numerical model - the battery was heated in the entire volume, instead of surface heating as in real experiment. In such a way the volume averaged temperature seems to be higher in the numerical simulation. However, despite this, both models almost perfectly agree in regard to the time moment, the battery cell fell into a chain reaction. This happened at about 75 s after heating begun. The chain reaction stopped when all reactants ran out. The heat release rate is shown in Fig. 4.

This confirms the violent chain reaction, which started to develop about 75 s of simulation. It lasted about 5 s and the peak power reached almost 14 kW. Then the reaction rapidly ceased in series of oscillations. The total energy released was 25.4 kJ (7.06 Wh).

In the next stage a module containing 9 cells was examined. A cell in the module corner (b_{11}) was initially damaged. The initial failure was modeled in a similar way as previously, but the heat flow provided to the first cell was limited to 15 W. It allowed for keeping more realistic circumstances: commonly a dozen of minutes pass in the damaged cell between an initiating event and the thermal runaway [23].

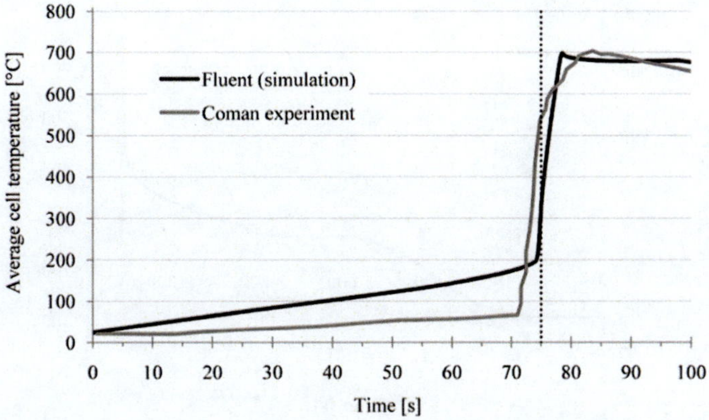

Fig. 3. Comparison of temperature increase between the presented model and Coman's experiment

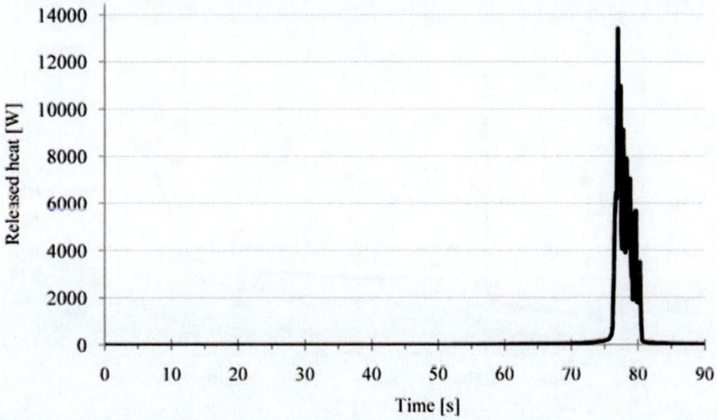

Fig. 4. Heat release rate.

The slower pace of temperature increase allowed also for better identification of sequence of triggered exothermic processes. The development of the thermal runaway in the first cell (b_{11}) is presented in Fig. 5.

The process started to be very violent at about 1080 s of simulation. The peak value of the released heat is slightly lower than previously due to slower process rate, but the total amount of released energy was similar (about 24 kJ = 6.67 Wh). The temperature in all cells in the module during the process of thermal runaway is shown in Fig. 6.

The first observation is that the other cells fell into the chain reaction almost in the same time. This happened at about 3220 s of simulation, which means the delay after the process took place in the first cell was about 2140 s.

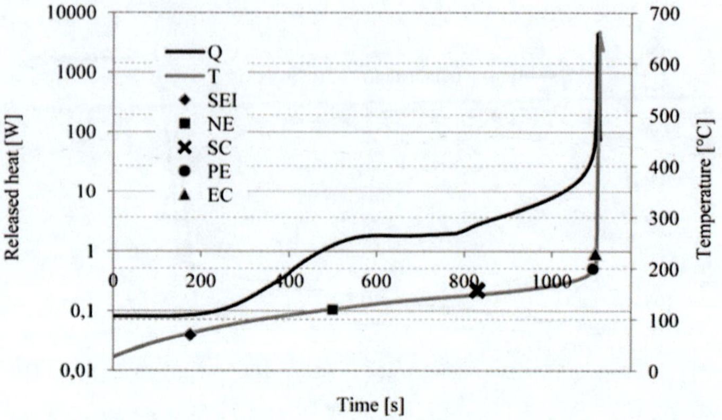

Fig. 5. The amount of released heat and temperature increase in the first cell (with the forced failure)

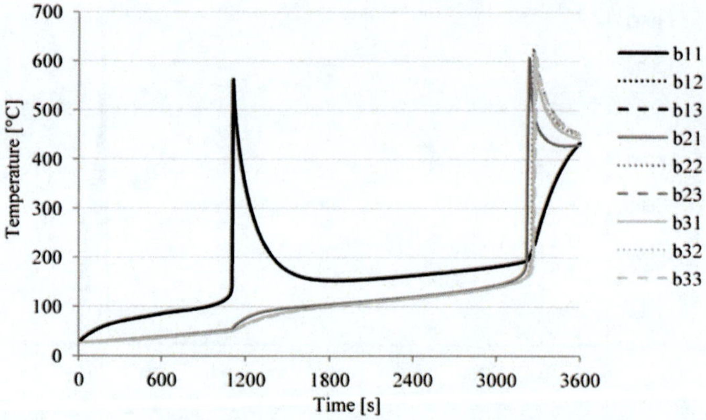

Fig. 6. The temperature change in all cells in the module

The determined time delay is generally in accordance with experimental investigations carried out by other researchers. For instance, Huang et al. examining experimentally different types of car batteries found the average value of this delay as 1715 s. Moreover, it strongly depended on battery type [23]. In this period heat from the first cell have been spreading over the module, triggering the exothermic processes in the neighboring cells. Figure 7 presents the zoomed time interval at about 3220 s of simulation, which allowed for determination of the exact sequence in which these cells were triggered.

The visible order of cells triggering is: at about 3220 s cells b_{12} and b_{21}, then cell b_{22} at 3255 s, next pairs b_{13} and b_{31} at 3260 s and b_{32} and b_{23} at 3270 s, finally b_{33} at 3275 s.

Each cell stores an amount of energy, which regardless of a particular process can be released if the thermal runaway occurs. When the chain reaction is completed this resource is exhausted. Figure 8 presents the development of the thermal runaway process in the entire 9 cell module.

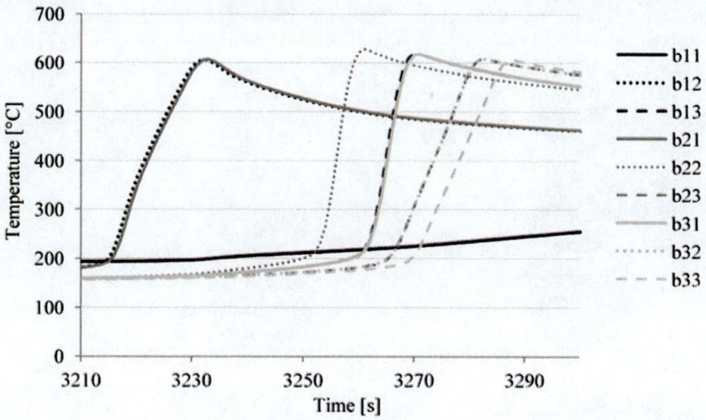

Fig. 7. The zoomed time interval in which other cells were triggered

What can be seen confirms the earlier observations. The chain reaction in b_{11} cell released the enough amounts of heat to warm up the module chassis and the neighboring cells, mainly cells b_{12} and b_{21}. These process lasted relatively long time (as was stated earlier - 2140 s). In these cells the chain reaction started at first (at 3220 s), the generated energy heat up cell b_{22} (at 3260 s) and at a moment later cells b_{31} and b_{13} (at 3270–3280 s). Cells b_{23}, b_{32} and at the end cell b_{33} fell into chain reaction soon (at 3280–3290 s). The amount of released heat was so high that this front of triggered cells would propagate over rest of the module with approximately the same pace. Finally, after 3290 s from the initial failure, the battery module would be likely to blow up and it would take about at most a dozen of seconds.

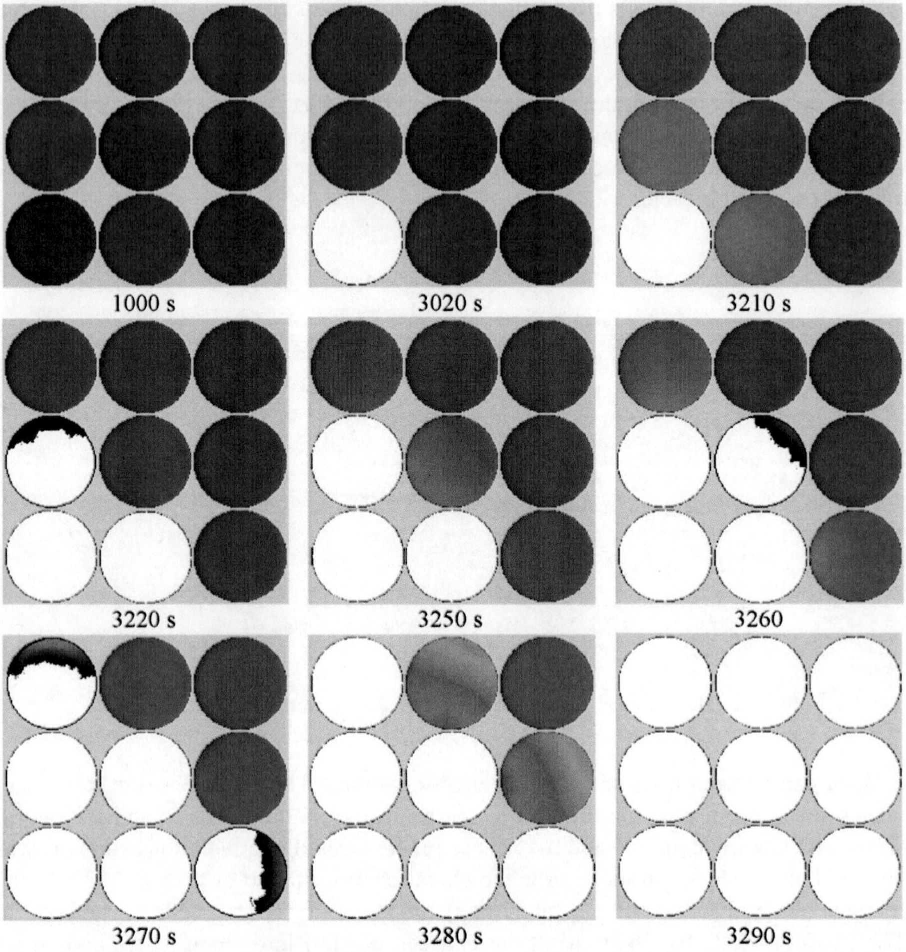

Fig. 8. The zoomed time interval in which other cells were triggered (grey level corresponds to the remaining energy)

5 Conclusions

The work presents a numerical research on the phenomenon of thermal runaway in a module of lithium-ion batteries. Commonly used cells of 18.650 type were examined. Despite there are a number of similar works, one should have in mind that the examined phenomena are very sensitive on details of a battery cell structure and the entire module construction. Therefore each assemble may behave in different manner, hence a new battery set should undergo detailed experimental and numerical tests to estimate the safety level during its usage.

At first, the correctness of the proposed numerical model of a single cell was proved by validating it with the literature experimental data. Next, the propagation of the thermal runaway over entire battery module was examined. It was shown, that such process may lead even to an explosion in a relatively long time after the occurrence of a primary cause. In the examined case the primer cell feel into chain reaction after about 18 min, the directly neighboring cells did this 36 min later. This wave reached the last group of cells in next 1 min. So, it took just about 55 min to cover the entire module. Having in mind the total amount of released energy one can imagine the seriousness of the threat. The latent nature of the described process might suggest some more dangerous scenarios: an unaware driver of an EV, in which a single battery cell starts to get out of control enters a tunnel or leaves his car in a underground parking.

The comparison of the obtained results with literature experimental data confirmed the model validity. This premise allows for applying the model for battery cells with different parameters.

The work in the future will be focused on the influence of different cooling methods on the rate of propagation of the thermal runaway wave inside a battery module. However, the reliability of such studies for a given battery type requires the knowledge of parameters of inner cell structure, which may sometimes be difficult due to manufacturers' reluctance to share data.

References

1. Poizot, P., Dolhem, F.: Clean energy new deal for a sustainable world: from non-CO2 generating energy sources to greener electrochemical storage devices. Energy Environ. Sci. **4**(6), 2003–2009 (2011). https://doi.org/10.1039/C0EE00731E
2. Aziz, M., Oda, T., Kashiwagi, T.: Extended utilization of electric vehicles and their re-used batteries to support the building energy management system. Energy Procedia **75**, 1938–1943 (2015). https://doi.org/10.1016/j.egypro.2015.07.226
3. Yuan, X., Li, L., Gou, H., Dong, T.: Energy and environmental impact of battery electric vehicle range in China. Appl. Energy **157**(1), 75–84 (2015). https://doi.org/10.1016/j.apenergy.2015.08.001
4. Wang, Q., Ping, P., Zhao, X., Jinhu, G.C., Chen, S.C.: Thermal runaway caused fire and explosion of lithium ion battery. J. Power Sour. **208**, 210–224 (2012). https://doi.org/10.1016/j.jpowsour.2012.02.038
5. Xia, L., Miao, H., Zhang, C., Chen, G.Z., Yuan, J.: Review-recent advances in non-aqueous liquid electrolytes containing fluorinated compounds for high energy density lithium-ion batteries. Energy Storage Mater. **38**, 542–570 (2021). https://doi.org/10.1016/j.ensm.2021.03.032
6. Zhang, L., et al.: Experimental study on the synergistic effect of gas extinguishing agents and water mist on suppressing lithium-ion battery fires. J. Energy Storage **32**, 101801 (2020). https://doi.org/10.1016/j.est.2020.101801
7. Broussely, M., et al.: Main aging mechanisms in Li ion batteries. J. Power Sour. **146**(1–2), 90–96 (2005). https://doi.org/10.1016/j.jpowsour.2005.03.172
8. Love, C.T., Baturina, O.A., Swider-Lyons, K.E.: Observation of lithium dendrites at ambient temperature and below. ECS Electrochem. Lett. **4**(2), 24–27 (2015). https://doi.org/10.1149/2.0041502eel
9. Liu, B., et al.: Safety issues caused by internal short circuits in lithium ion batteries. J. Mater. Chem. A **43**(6), 21475–21484 (2018). https://doi.org/10.1039/C8TA08997C

10. Spotnitz, R., Franklin, J.: Abuse behavior of high-power, lithium-ion cells. J. Power Sour. **113**(1), 81–100 (2003). https://doi.org/10.1016/S0378-7753(02)00488-3
11. Nishi, Y.: The development of lithium ion secondary batteries. Chem. Rec. **1**(5), 406–413 (2001). https://doi.org/10.1002/tcr.1024
12. Hendricks, C., Williard, N., Mathew, S., Pecht, M.: A failure modes, mechanisms, and effects analysis (FMMEA) of lithium-ion batteries. J. Power Sour. **297**(30), 113–120 (2015). https://doi.org/10.1016/j.jpowsour.2015.07.100
13. Isozumi, H., et al.: Application of modified styrene-butadiene-rubber-based latex binder to high-voltage operating LiCoO2 composite electrodes for lithium-ion batteries. J. Power Sour. **468**(31), 228332 (2020). https://doi.org/10.1016/j.jpowsour.2020.228332
14. Jia, Y., Uddin, M., Li, Y., Xu, J.: Thermal runaway propagation behavior within 18.650 lithium-ion battery packs: a modeling study. J. Energy Storage **31**, 101668 (2020). https://doi.org/10.1016/j.est.2020.101668
15. Zhang, W., Liang, Z., Yin, X., Ling, G.: Avoiding thermal runaway propagation of lithium-ion battery modules by using hybrid phase change material and liquid cooling. Appl. Therm. Eng. **184**, 116380 (2021). https://doi.org/10.1016/j.applthermaleng.2020.1116380
16. Ouyang, D., Weng, J., Hu, J., Chen, M., Huang, Q., Wang, J.: Experimental investigation of thermal failure propagation in typical lithium-ion battery modules. Thermochimica Act **676**, 205–213 (2019). https://doi.org/10.1016/j.tca.2019.05.002
17. Chen, M., Dongxu, O., Liu, J., Wang, J.: Investigation on thermal and fire propagation behaviors of multiple lithium-ion batteries within the package. Appl. Therm. Eng. **157**, 113750 (2019). https://doi.org/10.1016/j.applthermaleng.2019.113750
18. Zhang, L., Zhao, P., Xu, M., Wang, X.: Computational identification of the safety regime of Li-ion battery thermal runaway. Appl. Energy **261**, 114440 (2020). https://doi.org/10.1016/j.apenergy.2019.114440
19. Logan, E.R., et al.: A study of the physical properties of Li-ion battery electrolytes containing esters. J. Electrochem. Soc. **165**(2), 21–30 (2018). https://doi.org/10.1149/2.0271802jes
20. Coman, P.T., Darcy, E.C., Veje, C.T., Whited, R.E.: Modelling Li-ion cell thermal runaway triggered by an internal short circuit device using an efficiency factor and Arrhenius formulations. J. Electrochem. Soc. **164**(4), 587–593 (2017). https://doi.org/10.1149/2.0341704jes
21. Intano, W., Kaewpradap, A., Hirai, S., Masomtob, M.: Thermal investigation of cell arrangements for cylindrical battery with forced air-cooling strategy. J. Res. Appl. Mech. Eng. **8**(1), 11–21 (2020). https://doi.org/10.14456/jrame.2020.2
22. Kong, D., Wang, G., Ping, P., Wen, J.: Numerical investigation of thermal runaway behavior of lithium-ion batteries with different materials and heating conditions. Appl. Therm. Eng. **189**, 1116661 (2021). https://doi.org/10.1016/j.appthermaleng.2021.1116661
23. Huang, Z., et al.: Experimental investigation on thermal runaway propagation of large format lithium ion battery modules with two cathodes. Int. J. Heat Mass Transf. **172**, 121077 (2021). https://doi.org/10.1016/j.ijheatmasstransfer.2021.121077

Structure and Traffic Organization in Transport Systems

Impact of Drivers' Waiting Time on the Gap Acceptance at Median, Uncontrolled T-Intersections

Adrian Barchański(✉)

Faculty of Transport and Aviation Engineering,
Silesian University of Technology, Katowice, Poland
adrian.barchanski@polsl.pl

Abstract. The paper presents results of research on the influence of the waiting time of drivers of subordinate movement on the time duration of the accepted gap and the number of vehicles using it to continue their journey at selected median, uncontrolled T-intersections with two two-lane main roadways, located in the Upper Silesian agglomeration. Facilities located within and outside built-up areas were selected for analysis. A two-stage waiting model was proposed for drivers of subordinate movements. With regard to the analysis of the number of vehicles merging or crossing the traffic and follow-up time, it was proposed to use a new indicator. A qualitative and quantitative study was conducted in the characteristics of traffic conditions and driver behaviour at selected intersections. The results of tests conducted in the afternoon rush hours confirmed the hypothesis about the lack of influence of drivers' waiting time on the time duration of accepted gaps.

Keywords: Gap acceptance · Driver behaviour · Uncontrolled intersections · Linear regression

1 Introduction

Uncontrolled intersections are common facilities of the road transport infrastructure. They are located in all areas of the region's spatial layout, thus leads very diverse traffic [1]. Due to the strong influence of conflicting traffic streams at uncontrolled intersections, vehicles in subordinate movement suffer significant delays [1–3].

The continuous increase in the number of cars and their widespread use on a day-to-day basis makes the parameters of the road infrastructure insufficient and needs to be adapted to the current and anticipated needs of users. Nowadays investments in road infrastructure to improve traffic conditions are preceded by a planning and modelling phase to determine the appropriateness and desired direction for changing the current solution and to develop an optimal ones. When evaluating and selecting the best ones among potential solutions, commonly used methods are: analytical methods based on gap acceptance theory [4–6] as well as simulation methods based on the Wiedeman model.

E. Macioszek and G. Sierpiński (Eds.): TSTP 2021, LNITI, pp. 149–166, 2022.
https://doi.org/10.1007/978-3-030-93370-8_11

Incorrect planning decisions result in infrastructure solutions that are not enough match. This may result from tool inaccuracies. The cost of repairing infrastructure errors and costs for users far outweighs the planning costs. Accurately estimating the capacity of the point and line elements of the road infrastructure is all the more important as it forms the basis for estimating the capacity of the entire transport network. This in turn ensures the proper: spatial planning, strategic and operational decision-making in infrastructure management and traffic flows planning [7]. However, the methods indicated are as accurate as their algorithms, inputs, and parameter values.

Critical gaps and follow-up times are an important part of calibrating the model so that the real traffic conditions. There are mapped as closely as possible from the point of view of the purpose of the analysis. The values of these parameters remain constant during the analysis, which does not allow to take into account the impact of drivers' waiting times on the variability of their behaviour and their acceptance of gaps with an ever shorter duration than primary preferred. The existence of such a link has been identified in the literature. This means that drivers in subordinate movement accept gaps with a shorter time duration than in non-waiting conditions [8–12].

The question therefore arises whether the observed variability in the perception of the traffic situation at the intersection by individual drivers, as a waiting time increases, affects the change in the values of the critical gaps and follow-up times? Traffic saturation results in creating queues and increasing waiting times. Does it also increase the value of the critical gaps and follow-up times in comparison with the values estimated under the conditions of free movement and presented in the instructions? It should be emphasized here that the values of the critical gaps and follow-up times are determined on the basis of the time duration of gaps accepted and rejected by individual drivers.

The author attempted to investigate the impact of drivers' waiting times in subordinate movement on the duration of the gaps they accept and the number of vehicles entering from the queue using the same gap. The purpose of the survey is to determine whether the drivers' waiting time causes the change of values of accepted gaps. The results of the research conducted so far presented in the literature has not yet established whether this has an impact on global traffic conditions at intersections. The research was conducted in a systemic, global manner with regard to the intersection and not the preferences of individual drivers. Therefore, only the waiting time and the time duration of accepted gaps were recorded. Unlike before in the literature, where the process of changing driving decisions regarding the duration of accepted gaps was examined in detail by analyzing time duration of all gaps available to the individual driver, the order of occurrence of the gaps and the decision on acceptance as a function of waiting time. Confirmation of the correlation between waiting time and accepted gap duration would mean that the analytical and simulation methods used nowadays estimate the capacity of the relations and traffic conditions inaccurately in a given road-traffic situation and must be specified by introducing a dynamic change in the values of the critical gap and follow-up time when modelling traffic saturating at subordinate relations. To the best of the author's knowledge, this is the first study relating to this narrow issue. No similar article was found in the literature describing the conduct of research to the extent indicated.

In the article, information was collected on the factors influencing the variability of the duration of gaps accepted and rejected by drivers based on literature review. Median, uncontrolled T-intersections were presented, which are the object of the study. The proposed method for investigating the impact of waiting times on drivers' gap acceptance is discussed. The results of the surveys were presented on selected local objects.

2 Literature Review

The study and analysis of uncontrolled intersections has been ongoing for decades, but this topic remains relevant today. Further problems related to this issue are being solved and the description is becoming more detailed, precise e. g.: [3, 11, 13–16].

Continuing the classification adopted in the literature [17, 18], Table 1 summarizes the constant and variable factors influencing drivers' individual decisions to accept or reject a set of subsequent gaps.

Table 1. List of factors influencing the driver, affecting the duration of the accepted gaps.

Impact mode	
constant	Variable
Sex [8, 16, 17, 19–21] Age [8, 16, 17, 19–21] Individual drivers' characteristics [1–3, 16, 22–24, 26, 27] Driving style [16, 17, 24, 28] Cultural and regional features [17, 24, 28, 29] Technical parameters of the car (acceleration, length) [7, 8, 12, 24] Distance from the nearest oncoming vehicle of conflict stream and traffic flow [7, 11, 15, 21, 28] Experience [16, 20]	Time of day [1, 8, 9, 19] Temporary psychophysical conditions [16, 20] Waiting time, delays [2, 7–12] Queue length [8] Presence and number of passengers [8, 24] Weather conditions [24, 25] Lighting [24] Rainfall intensity [10] Air transparency [10] Pavement conditions [10]

Constant elements characterize an object, a region. While variables make critical gaps and follow-up times on an object additionally as a function of factors that occur temporarily at random, difficult to predict way. Their designation requires a very restrictive determination of the purpose, scope and time of the survey.

Studies of critical gaps, follow-up times and determining factors are very popular in developing countries [17, 24, 26, 28]. In India, special attention is paid to the specific local conditions and behaviour of individual drivers [17, 24, 28]. It has been shown that risk prone and overt, unpardonable extortion of priority depends primarily on individual, permanent personal characteristics and respect for: established, accepted rules and standards of conduct.

In [16] underlines that previous studies have devoted little space of the impact of individual characteristics of drivers and their driving characteristics on critical gaps and follow-up times. In addition, the [16, 20] works notes that the driving factors analysed so far are in fact correlated, remaining in a cause-and-effect relations, determined by additional variables that have not been studied in terms of the effect on critical gap and follow-up time. The differentiation of individual driver behaviour, risk propensity (determining higher driving speed) and perception of the seriousness of the effects of potential events determined by risky behaviour are influenced by: age, gender (indicated as the most relevant factor) but also other not empirically surveyed: driving experience both factual (positively correlated with speed, lane selection and general road behaviour) as well as perceived personality or stress levels as a temporary factor [16, 20].

As demonstrated, certain character traits are determined by others, so the cause and effect chain between actual independent variables and the duration of accepted and rejected gaps. It is multi-stage process. The most relevant from the point of view of the subject of the article proven correlations occur between age, gender, search for impressions and risky driving behaviour, higher speed, ignoring speed limits. However, even such extensive studies of the interdependence of personal characteristics and their impact on drivers' decisions did not take into account their impulsivity and aggressiveness [24, 26, 27].

However, as the authors of [10, 17] note, the results obtained can only be correct on a very narrow type of objects that were subject to analysis also because of the assumptions made and the way in which the results were get. One of them, which is very strongly determining the results obtained and which seems not to be in line with reality, is the consideration of available gaps in the major stream by drivers so as not to force priority. While there are two-way interactions between conflicting streams and the speed of vehicles of superior movement is not constant, these drivers can slow down to facilitate the enter or merge of other drivers (especially in traffic saturation conditions) [30] or to avoid collisions [21, 24]. The driver of the subordinate movement can accelerate for the same reasons. The process of entering or merging the traffic affects the distribution of gaps in the superior traffic flow [21, 30].

In the actual traffic situation, the behaviour of subordinate and superior drivers varies according to the degree of saturation of traffic. There are cases where the order of traffic at intersections is not due to traffic regulations and organisation, but to various forms of social behaviour and the types of vehicles interacting with each other [1]. Transitions between the two states highlighted are often observed in real-world conditions, most often during peak hours, of high congestion, while many models used to describe traffic allow for limited mapping of identified the types of situations [30].

The results of the [8, 9] study confirmed that drivers after longer waiting time to accept shorter gaps than they had originally assumed. Of course, there is also an inverse dependency on variables. Well, shorter gaps in the superior traffic flow make the wait time longer. In addition, drivers' behaviour in this area changes over the course of the day. In the later hours, they are more likely to risk than in the morning.

The impact of waiting times on drivers' increased risk propensity and acceptance of shorter gaps has also been confirmed in the works [11, 12, 16, 21]. In [12], however, a weak negative correlation between the waiting time and the critical gap was marked and not directly measured in the sample of the duration of the accepted gaps. The results at work [10] are different, where only a negligible effect of the waiting time of first-time drivers in the queue on the critical gap has been shown.

3 Median, Uncontrolled T-Intersections

Despite a wide range of studies conducted at uncontrolled intersections little space has been devoted to median, T-intersections with a two two-lane, one-way carriageways of major road, in particular as regards the impact of external factors on critical gap and follow-up time. It is therefore necessary to provide a brief description of the type of facilities to be tested here.

Features and differences from other types of uncontrolled intersections are due to geometry and number of approaches [4, 18, 31, 32]. The major road plays an important role in the road network of the region, the city of agglomeration, is assigned to high-class roads, since only such roads have two two-lane, single-way carriageways. This significantly increases capacity, handles very large traffic flow and the distribution of them into two lanes affects the magnitude of the priority stream [18, 31, 32]. In addition, drivers on the major road drive through the intersection without delays, more smoothly, than if the object had been equipped with traffic lights. Its role is to facilitate the incorporation of subordinate movements into traffic while increasing the safety of these manoeuvres [32]. The geometry of the intersection of the type, together with an indication of the permitted movement, the hierarchy of streams and the trajectory of the movement, is shown in Fig. 1.

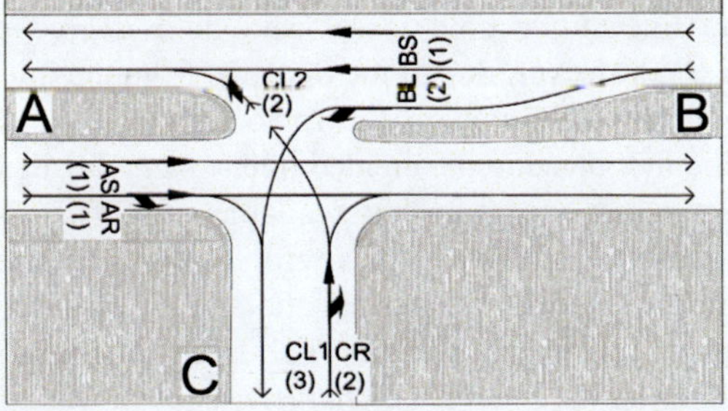

Fig. 1. Geometry and trajectory of movements at median, uncontrolled T-intersection. Numbers in brackets show hierarchy order

On the facilities of the objects, the use of traffic lights is not necessary, since the volume of traffic on the minor road is small in relation to the size of the streams on the major road. The increase in safety and facilitation of movements was achieved through the use of a wide, median strip, thus creating a space where vehicles waiting for the possibility of continuing to drive are stopped at the intersection [4, 18, 32, 33]. The risk of rear-end crashed caused by a vehicle turning left from the major road is also reduced. The left turn movement from the minor road can be performed in two independent stages. The first one (CL1) is related to the entrance from the minor approach and the stop in the area of the wide, median strip, and the second one (CL2) involves entering traffic starting from the wide median strip. The presence of only one minor approach means that there are no complex impedance effects between different order streams. The other two subordinate movements include a left turn from the major road (BL) and a right turn from the minor road (CR).

The geometric features described clearly determine the advantages of objects of this type [4, 18, 32, 33]. These include the aforementioned increased safety, efficient operation of all user groups, high capacity through impact on volume of conflicting streams and two two-lane major road carriages, channelization, separation of streams, unambiguous location of collision points at the intersection and reduction of their area, reduction of the number of decisions need to be made at the same time. The characteristics of the objects of the type different from other types of intersections are already revealed when analysing the hierarchy of movements, their interdependencies and the way in which individual manoeuvres are performed. The characteristics of median, uncontrolled T-intersection in terms of impact on the critical gaps and follow-up times are shown in the [18].

4 Method of Analysis of Impact of Waiting Time on Gap Acceptance

On the basis of the literature analysis and observations made, the assumption was made about a two-step process of waiting of drivers of subordinate movement. In the case of a permanent queue of vehicles, with a large number of cars and high volume of superior streams, it is assumed that drivers waiting at the end of the queue and moving gradually to the stop line suffer time losses and their degree of nervousness and aggressiveness increases, however, due to the distance from the intersection dial are not able to analyze the current situation on the major road on an ongoing basis. Only drivers who occupy the first place in the created queue can realistically assess the traffic conditions, decide to reject or accept further available gaps. This is the second stage of waiting process, during which drivers strive for a fast but safe execution of the manoeuvre. It is therefore necessary to assess the situation soberly, regardless of the time spent inside the queue. Drivers will take a risk-free decision to get involved, with a degree of risk depending on the individual characteristics of the driver, their skills and the characteristics of the vehicle in local conditions. These are revealed in the decision to accept a gap with a specific duration as a function of waiting time.

As part of the research, the gap was defined as the time interval between the passage of two successive vehicles of the superior stream through the determined cross-section of the carriageway on the major road, determined by the axis of the lane of the subordinate relation. The gap duration is the time when in the cross-section defined on the carriageway of the major road there is no part of the body of two successive vehicles of the major stream. The gap is called accepted when it is used to merge or crossing the traffic or is called rejected when the driver of the subordinate movement, despite occurrence of the gap, refrains from continuing the trip. It should be stressed that, in the course of the survey, the waiting time of drivers was measured as the time elapsing between the occurrence in the superior stream of two consecutive gaps accepted by the drivers of the subordinate movement. However, the time value sought is the difference between the moment when the gap accepted by the driver starts and the moment when the previous gap accepted by the vehicle crossing or merging the traffic ended. In addition, the duration of the accepted gap and the number of vehicles using it were recorded for each such situation. The analysis proposes to use a new indicator of gap duration per vehicle ($W_{d/v}$) designated as the quotient of the time duration of the gap accepted by at least two consecutive drivers in a given subordinate movement and the number of drivers accepting the gap.

Correlation and linear regression analysis was used to identify and assess the force of the impact of waiting time on drivers' behaviour in subordinate movement. If the results are confirmed and the statistical significance of the difference between the correlation coefficient and the slope from zero is verified, the effect of waiting times on driver behaviour can be said to occur.

Two separate studies were carried out. In the first, the duration of the gap accepted by one driver of the subordinate relationship was assumed as a dependent variable. In the second, the dependent variable was the designated indicator ($W_{d/v}$). The results of the survey will answer two question as follows:

- whether the longer waiting time at the subordinate movement for the occurrence of an accepted gap increases the risk propensity of drivers and makes them to decide to enter traffic in gaps with a shorter duration?
- whether subsequent vehicles enter from the subordinate movement in a shorter interval than in situations where these manoeuvres are carried out without or with a shorter waiting time?

The adoption of the waiting time as the time elapsing between two successive gaps in the major stream and the examination of the entry of subsequent vehicles of subordinated movement with the use of the proposed indicator is an approach that has not been given in the literature much attention so far.

A detailed formal description and rules for the application of correlation and regression analysis are given in [34, 35]. It is a tool often used in scientific research not only in traffic analysis but also in the survey of gap duration variability [12, 14, 16, 20, 21]. In the interests of clarity, the terms and methods adopted and used in the article were decided here for short description of them.

Correlation examines in both qualitative (scatter diagram) and quantitative (correlation coefficient) the existence of relations between variables, testing the significance of the results obtained for the assumed hypotheses and interpreting the results. Information on the strength and direction of variable interdependence is sought here and the systematic nature of interdependence is attempted [34, 35]. While regression analysis allows you to quantifically describe the relations between variables to analyze its structure, determine the significance of a random factor, and as a result, predict based on a built model [34, 35].

Correlation and regression analysis are based on a number of assumptions, the fulfilment of which by the collected measurement data is a prerequisite for reliable inference based on the results of empirical studies and a built, calibrated model. The criterion of conformity, correctness of the model is a broad, multicriterial analysis of the residuum, the order of actions is presented in the diagram (Fig. 2).

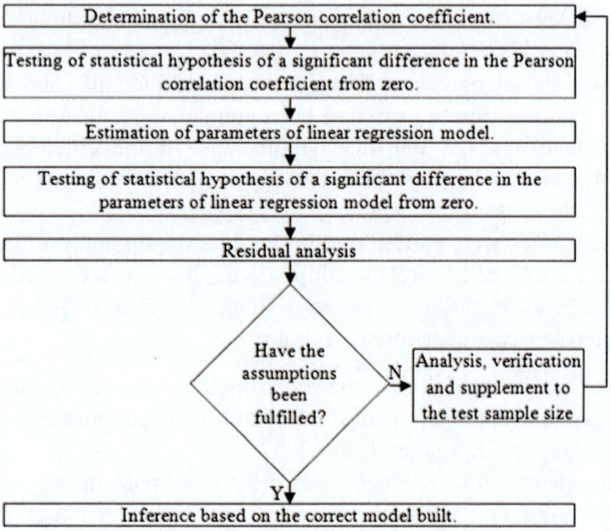

Fig. 2. Workflow for correlation and regression analysis

The first stage of the study involves analysing the collected data, drawing up a correlation and identifying possible outliers. The analysis of the relations between the variables was examined using Pearson correlation coefficient r_{xy} and linear regression. It was accepted in the study that the lack of correlation is said when $|r_{xy}| < 0.\ 4$, with a strong correlation $|r_{xy}| > 0.\ 7$ in other cases and with a weak correlation [35]. Statistics with n-2 degrees of freedom (where: n - sample size) [34, 35] were used to verify that the designated r_{xy} value is significantly greater than zero.

The regression analysis was then carried out, i.e. determining the strength of the relations between the variables, building a functional relations that maps the interdependence of the two characteristics. A linear model of $y = b_1x + b_0$ with one independent variable is adopted. The function was matched to the measurement data using the least squares method [34, 35].

However, it is necessary to examine whether the measures set reliably describe the course and shape of the dependencies. It should be checked whether the determined parameter values are statistically significantly different from zero. The statistics used in the verification of compliance with the assumptions was based on the following literature [34, 35, 38, 39].

After building the model and verifying its significance, you can use residuum to go to the last stage of analysis. That is, to verify that three assumptions are met that determine the possibility of later inference and prediction on the basis of the built model. To verify the hypothesis that the residuum of the model has a normal distribution, the Shapiro-Wilk test with U_n test statistics was used. The Goldfeld-Quandt test with Sj^2 statistics was used to check the variance constancy of a random component, or uniform dispersion of data along the regression line. In contrast, the Durbin-Watson test, with statistics labeled *d*, was used to examine the autocorrelation of the residuum of the model. Only after all four assumptions have been met can the built model be used for inference and forecasting.

5 Characteristics of Research Objects

Empirical studies were carried out on two selected intersections located in the area of the Upper Silesian agglomeration. Their location is shown in Fig. 3.

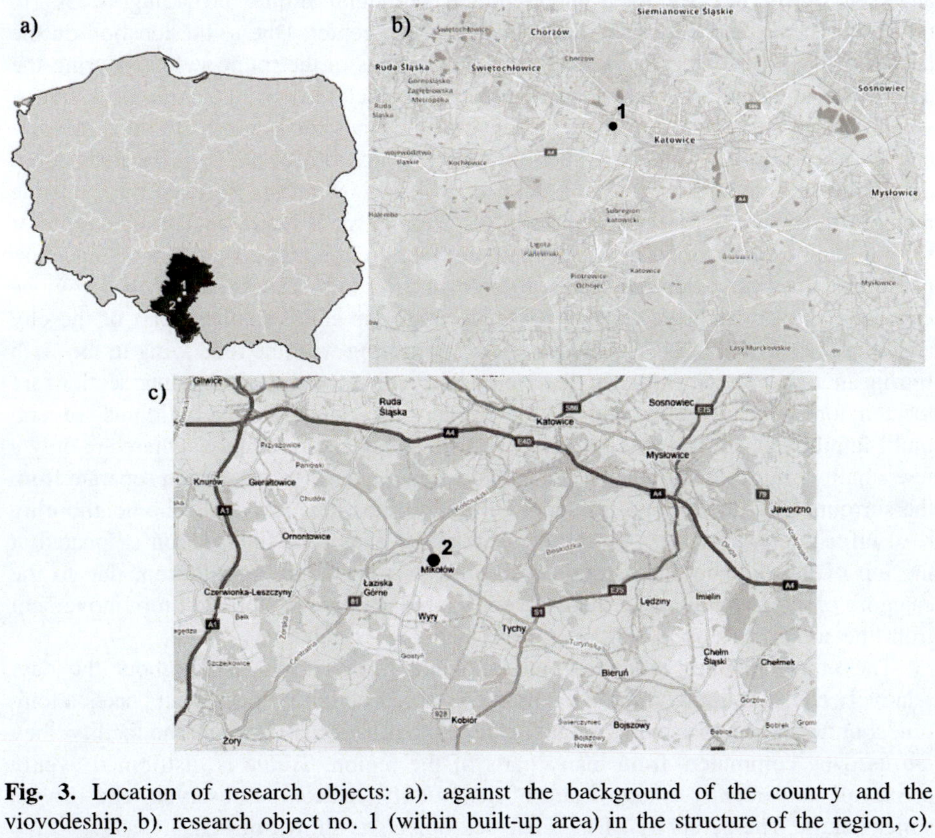

Fig. 3. Location of research objects: a). against the background of the country and the viovodeship, b). research object no. 1 (within built-up area) in the structure of the region, c). research object no. 2 (outsider built-up area) in the structure of the region (Source: Own research based on [36, 37])

The first of the studied intersection (research object no. 1) is located in a built-up area on the border of one of the central districts of Katowice city. The major road is an important, one of the few connections between the north and south of the city and the inner city area. Thanks to its meridial course, it connects neighbouring cities in the center of the agglomeration. On the other hand, the minor road plays an important role in the area of the district, allowing access to housing estates, workplaces, businesses and downtown. The direct functional area of the property performs a residential and service role. The object is separated from the environment through the course in the ditch, both roads are straight in plan and profile. There is no influence of pedestrians and bus stops. Due to the location of the facility in the center and the dense network of urban roads, there is an interaction of nearby intersections with traffic lights. Equal traffic flows in all movements during rush hour at nearby intersections is regulated by traffic lights. This, in combination with the distance of several hundred meters from the local object, means that there is no platooning of vehicles in the major stream. There is constant, regular flow with a uniform intensity determined by the signaling phases. The analysed object handles local inbound and outbound traffic, providing access to recreation, health and office facilities located in the center. Due to the location on the border of the city center, there are strong fluctuations in the traffic volume during the day. There is a sharp increase in the number of vehicles reporting during peak hours, which causes temporary congestion, long waiting times for subordinate movements.

The second of the analysed intersections (research object no. 2) is located within the administrative boundaries of the city of Mikołów (Silesian Voivodeship), outside the built-up area. The object lies at the intersection of roads important from the point of view of the whole region. The major road provides an important, significant connection between industrial centers and cities located in the south-eastern part of the province and beyond. The subordinate road passes through the entire southern part of the city and continues in a south-westerly direction, connecting with the road lying in the sixth European transport corridor. In the immediate surroundings of the intersection are located forest areas, green, and a little further low, single-family buildings, several multi-family houses. The area primarily performs residential functions, there are only a few small workplaces, storage areas. The location makes the intersection separate from the surroundings, there is no influence of: pedestrians, bus stops and the neighboring road infrastructure facilities on the distribution of gaps. The intersection is located at the top of the hill, on a horizontal arch with a large radius. In addition, due to the category of intersecting roads, there are entry auxiliary lane for the right turn movement from the subordinate road and the second stage of the left turn.

The object handles inbound, outbound and throw traffic. Throughout the day, relatively constant traffic volume is observed, increased during peak hours takes a long time, but the increase is small, gradual. Due to the role of road forming the facility, they are used by commuters from many parts of the region. Traffic is distributed evenly across all movements, a large share of heavy-duty vehicles is observed, and drivers suffer delays. The geometry of the analysed objects is shown in Fig. 4.

The survey was conducted in spring 2018 on representative working days, during the afternoon rush hour. The traffic situation was recorded using cameras located at a height of about 4 m above the intersection surface, Footage of 4 days of recordings for each object, recording the image at a rate of 24 frames per second allowed to determine the duration of accepted gaps and waiting time. The recordings were analyzed frame by frame with an accuracy of 0.042 s using the Black Magic DaVinci Resolve 14 program.

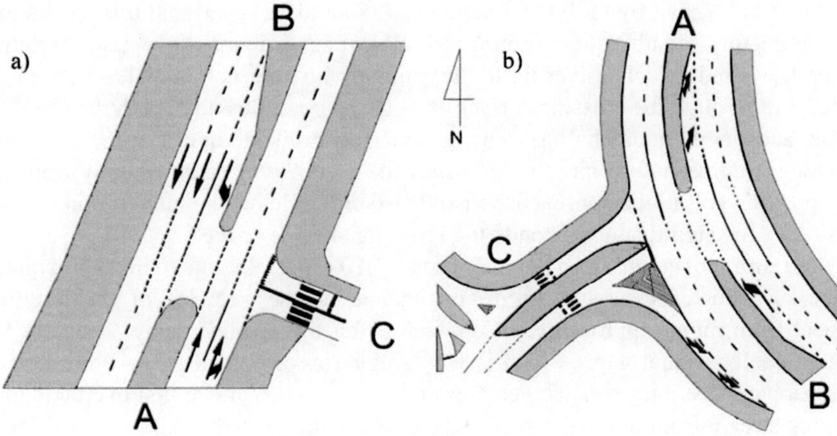

Fig. 4. Geometry of analyzed intersections: a) object no. 1, b) object no. 2

On research object no. 1, the throw traffic on the major road towards the centre (AS) is predominating. It has the greatest impact on the delays in the left turn movement of the subordinate approach (CL1), causing only a dozen vehicles per hour to continue their journey. The nature of traffic is a rapidly changing process, Just over 10% of vehicles turning left from the minor approach (CL1) merge the intersection without stopping. Slightly lower volumes occur on the major road in the opposite direction (BS), making it easier for vehicles of the second stage of the left turn (CL2) to turn left, also because only the priority stream in the left lane of major road interacts. Due to the long waiting time for entering the traffic, the intensity of subordinate relation is incomparably less than the already indicated priority stream traffic volume. The number of vehicles entering from the subordinate approach was strongly influenced by vehicles turning right from the major road (CR), especially since the share of this movement in the volume of all turning relations was predominant.

The accumulation area in the wide median strip accommodating only one vehicle under conditions of increased traffic also made it difficult to perform the manoeuvre, since the occurrence of a sufficient gap at the first stage of left turning from minor approach (CL1) was not used due to the lack of space to stop on the occupied accumulation area at the intersection.

At the object no 2, i.e. located outside built-up areas, a left turn queue (BL) was continuously recorded during periods of high volume of traffic, but the length of the auxiliary lane was sufficient as an accumulation area, so that the vehicles waiting for the left turn from major road did not block the lanes intended for straight movement (BS). However, the impact on the straight movement on major road is noticeable. By reducing the speed and manoeuvring of vehicles of this subordinate movement (BL), the left lane of major road is reluctantly used by first order vehicles. It is occupied only by passenger cars, also for purpose of overtaking the columns of heavy vehicles.

In this left turn from major road movement (BL), however, waiting times vary greatly. This means that entering or crossing the traffic is carried out more often by groups of vehicles than by individual vehicles. It should be noted that this results in a different trajectory of subsequent turning vehicles of subordinate movement. The first, the longest-awaited vehicle drives far to the centre of the intersection dial to shorten the potential path through the crossing surface as much as possible and exploit the smallest, available, acceptable gap. The last vehicles entering from the queue at the same gap move along a curve with a large radius, start the manoeuvre early, directly from the waiting position still in the approach area and also aim to leave the intersection as soon as possible. This significantly extends the crossing surface space.

The left turn movement from the major road (BL) especially due to the high volume of traffic of vehicles causes a strong impedance of the vehicles of the left turn movement from minor approach (CL1), which suffer the greatest delay. They are left with only small residual gaps, which is why single, less frequent group entries dominate. It happens that long-awaited vehicles at the opportunity move first overtaking the left turning stream from the major road (BL) or, depending on the skills of the drivers, bypass the vehicles waiting for the intersection shield, taking advantage of the gaps rejected by other drivers.

Vehicles in the second stage of the left turn from minor road (CL2) due to low traffic volume in the left lane on the major road enter the traffic very smoothly. Drivers of this movement suffer a small delays and easily enter the traffic in the priority stream also by equating the speed using the available auxiliary lane. On many occasions, these drivers did not use the available auxiliary lane at all, but at low traffic volume of major stream they immediately drove into the main lane.

The situation in the right turn movement from the minor approach (CR) is slightly different. The high volume of traffic of vehicles entering rather individually in combination with the large volume of the priority stream (AS) makes the auxiliary lane used in its entirety, and the entering take place synchronously, very smoothly into the small available gaps in the major stream. Sometimes it was necessary to stop the vehicles of this subordinate movement in anticipation of a sufficiently long gap.

During peak hours, there was an even increase in traffic volume on all movements, not abrupt, as large traffic volumes are constantly recorded. The shares of the vehicles types and directional structures are mainly affected.

The tested objects are separated from the surroundings, The analysis was carried out in sunny weather, on representative days, in a homogeneous period. The waiting time determined by the traffic conditions at the intersection was the only observed factor influencing the decision to accept the available gaps. The situations selected for the analysis met the assumptions regarding the conditions by the observed traffic [31, 32]. During the analysis, there was no influence of other factors identified in the literature or described in the capacity estimation manuals.

6 Survey Results

After presenting the traffic characteristics on the analysed objects during the afternoon rush hour in a qualitative way, you can proceed to quantitative analysis. Table 2 shows the results of correlation and regression analysis of the impact of the waiting time of the first vehicle in the queue as an independent variable to: the length of the gap accepted by one driver (analysis 1) and the value of the proposed $W_{d/v}$ indicator, which is the ratio of the duration of the gap in the priority stream accepted by at least two vehicles and the number of vehicles using it (analysis 2). The survey was conducted on two local intersections located in built-up areas and outside built-up area.

Table 2. Results of the analysis of the correlation and regression between waiting time and driver behaviour for analysed local objects.

Survey	Analysis 1						Analysis 2					
Object	1	1	1	1	2	2	1	1	1	1	2	2
Movement	CL2	CL1	BL	CR	CL1	BL	CL2	CL1	BL	CR	CL1	BL
Sample size	126	219	92	127	127	142	419	322	215	161	201	192
r_{xy}	−0.076	−0.035	0.138	−0.018	−0.016	0.018	−0.141	−0.181	0.082	0.040	−0.014	0.010
$Se(r_{xy})$	0.090	0.068	0.104	0.089	0.089	0.085	0.048	0.055	0.068	0.079	0.071	0.073
t_r	−0.848	−0.520	1.324	−0.197	−0.183	0.214	−2.901	−3.286	1.198	0.505	−0.192	0.143
p-value	0.398	0.603	0.189	0.844	0.855	0.831	0.069	0.113	0.232	0.614	0.848	0.886
b_0	5.326	7.663	5.874	6.079	6.159	6.505	4.718	7.471	5.672	6.116	4.725	5.736
$Se(b_0)$	0.089	0.068	0.104	0.089	0.068	0.084	0.049	0.056	0.068	0.068	0.071	0.072
t_{b_0}	31.62	32.33	16.17	36.61	30.31	32.93	46.43	33.68	26.91	30.37	23.56	29.08
p-value	0.000	0.000	0.000	0.000	0.000	0.000	0.000	0.000	0.000	0.000	0.000	0.000
b_1	−0.011	−0.003	0.021	−0.003	−0.002	0.001	−0.011	−0.016	0.009	0.006	−0.001	0.001
$Se(b_1)$	0.013	0.006	0.016	0.013	0.012	0.004	0.093	0.088	0.092	0.074	0.015	0.005
t_{b_1}	0.013	−0.520	1.324	−0.197	−0.183	0.214	−2.901	−3.286	1.198	0.505	−0.192	0.143
p-value	0.398	0.603	0.189	0.844	0.855	0.831	0.004	0.001	0.232	0.614	0.848	0.886
U_m	0.986	0.979	0.971	0.984	0.989	0.989	0.970	0.985	0.989	0.977	0.973	0.972
p-value	0.228	0.086	0.077	0.201	0.305	0.300	0.164	0.198	0.302	0.090	0.086	0.080
S_j^2	1.337	1.140	1.100	1.442	1.524	1.275	1.146	1.114	1.162	1.187	1.394	1.133
$F_{critical}$	2.000	1.374	1.642	1.524	1.910	1.486	1.257	1.298	1.378	1.451	1.631	1.404
Residuum b_1	0.002	0.003	0.003	0.004	−0.004	−0.030	0.000	0.001	−0.003	0.000	0.001	−0.030
$t_{b_{1Residuum}}$	0.598	1.435	0.544	2.245	−1.330	−1.379	−0.530	0.381	−1.658	0.050	0.490	−1.637
p-value	0.551	0.153	0.588	0.127	0.186	0.392	0.596	0.703	0.099	0.961	0.625	0.599
d	1.834	1.946	2.139	1.761	1.947	1.849	1.878	1.992	2.001	1.869	1.980	1.855
d_l	1.670	1.765	1.670	1.670	1.670	1.710	1.835	1.805	1.750	1.710	1.750	1.750
d_g	1.740	1.795	1.700	1.740	1.740	1.760	1.850	1.830	1.790	1.760	1.790	1.790

Where:

r_{xy} – Pearson correlation coefficient
$Se(r_{xy})$ – standard error of Pearson correlation coefficient r_{xy}
t_r – test statistic on t-Student distribution for r_{xy} coefficient
b_0 – y-intercept of linear regression
$Se(b_0)$ – standard error of regression coefficient b_0
t_{b0} – test statistic on t-Student distribution for y-intercept of linear regression (b_0)
b_1 – slope of linear regression (b_1)
$Se(b_1)$ – standard error of regression coefficient b_1
t_{b1} – test statistic on t-Student distribution for slope of linear regression (b_1)
U_m – Shapiro-Wilk test statistic
S_j^2 – Goldfeld-Quandt test statistic
$F_{Critical}$ – critical value in Goldfeld-Quandt test
Residuum b_1 – the slope of the linear regression of residuum
t_{b1} residuum – test statistic on t-Student distribution for slope of linear regression of residuum
d – Durbin-Watson test statistic
d_l – lower bound of Durbin-Watson test
d_g – upper bound of Durbin-Watson test

The literature indicates that the correlation and regression analysis requires a sample of at least 30 items, the recommended size is about 100 items. In the case of larger samples, it is recommended to divide into subgroups and analyze independently in each of them [34, 35, 38, 39]. For each subordinate movement, all assumptions were met in both analyzes, which allows the constructed model to be used for research and inference. Thus, it was confirmed that the collected research sample is sufficient. In each study, the results of the verification of statistical hypotheses for individual tests were consistent with the assumptions of the correlation and regression analysis [34, 35, 38, 39].

At the intersection located in the built-up area, all four subordinate movements were examined. On the other hand, due to the presence of auxiliary lane at an intersection located outside the built-up area for subordinate movements: right turn (CR) and the second stage of left turn (CL2), the drivers did not wait for the manoeuvre to be carried out, so the method proposed in this case could not be applied. Accordingly, only the driver behaviour in the left turn movement from the main road (BL) and the first stage of the left turn from minor approach (CL1) was examined. The results obtained clearly indicate that there is no statistically significant correlation between the variables analysed. The model was validate based on four defined assumptions. These include the non-contradiction of the rest of the model with the normal distribution, the absence of heteroskedasticity in the recorded sample, the absence of a trend for the variability of the residuum of the model during observation, and the absence of autocorrelation of the residuum. These conditions for all analyses have been met. The results thus support the initial hypothesis that there is a hierarchy of factors determining the critical gaps and follow-up times, which has not yet been sufficiently examined. In the case of analysed intersections, the waiting time between the occurrence of two consecutive accepted gaps do not affect the behaviour of drivers. The same results were obtained for the all examined objects, which are located both within built-up areas and outside built-up areas. This is linked to the pursuit of the manoeuvre primarily in a safe and compliant manner at the expense of extending travel time. Drivers prefer to perform the maneuver in a safe manner, resigning from the risk and acceleration of the trip (making up for the delay associated with excessive waiting time), which would manifest itself in accepting shorter and shorter available gaps along with the increase in waiting time.

Assumptions are also an important element in the studies carried out. It should be noted here that the analysis included a homogeneous time period, only an afternoon rush hours. Previous studies in the literature have not been so restrictive in the selection of the analysis period.

Contrary to previous studies presented in Sect. 2 and to popular opinion, a longer waiting time for the possibility to continue driving in the afternoon rush hour does not affect the acceptance of shorter gaps in the major stream. This is true when the analysis is performed globally, without analyzing the course and variability of the decision making process by individual drivers. Only drivers in the first place in the queue can reliably observe the priority streams, assess the volume size and speed of the vehicles of the priority stream, the distribution of the duration of the available gaps and the possibility of continuing the trip depending on their own preferences. While drivers in the next positions in the queue cannot observe the current traffic situation, they have to wait for the manoeuvre to be carried out by the preceding vehicles. Although the length of the waiting time certainly increases nervousness, however, it is only the confrontation of the available gaps with one's own skills that affects how long the gap will be necessary to be able to continue driving.

The variability in the duration of rejected gaps by individual drivers with increasing expectations was not recorded, but the phenomenon was analyzed globally. Only the duration of the accepted gaps depending on the waiting time was investigated. The results confirmed that, regardless of the length of the queue of vehicles in subordinate movement and the waiting time for the accepted gap, they could be used critical gaps and follow-up times values characterising the survey object designated in the conditions of no queue to analyse capacity and assess traffic conditions. There was no statistically significant impact of waiting times on the variability in the time duration of accepted gaps.

The article presents studies not yet carried out in the literature. An attempt was made to answer the question of whether the length of the waiting time makes the gap in the priority stream of a certain duration accept a different number of vehicles and the interval between successive vehicles entering in turn will vary. In this regard too, the results of the study indicate that, at the afternoon rush hours, longer waiting times do not cause subsequent vehicles accepting the same gap to be involved in traffic at shorter intervals than in the case of shorter waiting times or lack thereof. The results obtained are independent of the area in which the intersection is located. Because the same results were obtained for all tested objects, which were located both within built-up areas and outside built-up.

7 Summary

The article presents the results of studies on the impact of the waiting time of the first-in-line vehicle in all subordinate movement on the behaviour of drivers and their propensity to risk. It manifests itself in accepting a gap of a certain duration and changing the number of vehicles using the same gap duration in the priority stream.

The analysis covered traffic during typical working days, in the afternoon rush hours at median, uncontrolled T-intersections with two two-lane major road carriageways located both within built-up areas and outside built-up. The studies used correlation and linear regression analysis together with a full assessment of the statistical significance of the results obtained. Verification included assessment compliance of the data with the assumptions of the method used. Based on the literature review, a set of factors were identified that influence the variability of the duration of accepted gaps.

It has been shown that on the examined objects, the effect of the elapsed, waiting time between the occurrence of two consecutive accepted gaps in the priority stream is not statistically significant in all subordinate movements regardless of the environment at which the intersection is located. This means that the length of the waiting time does not cause drivers to decide to continue driving by accepting shorter gaps. This means that the critical gaps and follow-up times describing the traffic at the intersection under study do not change with the increase of the waiting time. Moreover, if several vehicles enter during the same gap in the priority stream, a longer waiting time does not cause them to cross the stop line at shorter intervals, which would explicitly be associated with allowing a higher level of risk. This also confirms the hypothesis in the article that there is a hierarchy of factors determining the acceptance of gaps.

It is also important that this article analyses for the first time the effect of the factor on critical gap and follow-up time values only during a homogeneous period, i.e. during afternoon rush hour. On the basis of the results obtained, a hypothesis may be, drivers are aware of the increased volume of traffic requiring full concentration, attention, focus and are willing to accept a higher level of risk throughout their trip. This may be related to the specific role of the type of object being tested on the road network. High volume of priority streams, high speeds and two lanes on the major road make it the geometry of the intersection, the type of manoeuvre performed are the main determinants of the variability of critical gaps and follow-up times. It is related to the geometrical features of the objects of the examined type and the organization of traffic described in point 3, which occur in the examined objects, but the variability of the influence of these additional factors mentioned above was not investigated during the works on this article. As the reason for the lack of influence of waiting time on the gap acceptance. However, this hypothesis requires further research and verification among others by performing tests in the morning rush hour, in the remaining periods of the day, and then comparing the obtained results. Furthermore, it has not yet been examined whether there is a causal relationship between the identified factors determining the duration of the gaps accepted and rejected and, if so, whether it is multi-stage, to what extent and between which factors.

Further, extensive research using the proposed method is therefore necessary in order to be able to extend the inference of captured trends to a given type of intersection and to verify that studies carried out in the same way on other types of objects also show that the factor in question is not affected.

References

1. Prasetijo, J., et al.: Capacity analysis of priority intersections with flare under mixed traffic conditions. Procedia Soc. Behav. Sci. **138**, 660–670 (2014)
2. Kearney, J.K., Grechkin, T., Cremer, J., Plamert, J.: Traffic generation for studies of gap acceptance, pp. 177–186 (2006). https://citeseerx.ist.psu.edu/viewdoc/download?doi=10.1.1.122.6255&rep=rep1&type=pdf
3. Ashworth, R.: The capacity of priority-type intersections with a non-uniform distribution of critical acceptance gaps. Transp. Res. **3**(2), 273–278 (1969)
4. Transportation Research Board: Highway Capacity Manual, National Research Council, Washington, DC (2000)
5. HBS. Handbuch für die Bemessung von Strassenverkehrsanlagen. Forschungsgesellschaft für Straßen- und Verkehrswesen e. V., Köln (2001)
6. GDDKiA. MOP SBS. Metoda obliczania przepustowości skrzyżowań bez sygnalizacji świetlnej. GDDKiA, Warszawa (2004)
7. Beanland, V., Lenné, M.G., Candappa, N., Corben, B.: Gap acceptance at stop-controlled T-intersections in a simulated rural environment. Transp. Res. Part F. **20**, 80–89 (2013)
8. Nabaee, S., Moore, D., Hurwitz, D.: Revisiting driver behavior at unsignalized intersections: time of day implications for two-way left turn lanes (TWLTL) In: Proceedings of the Sixth

International Driving Symposium on Human Factors in Driver Assessment, Training and Vehicle Design, Olympic Valley - Lake Tahoe, California. Iowa City, IA, 27–30 June 2011. Public Policy Center, University of Iowa, pp. 440–446 (2011)
9. Hamed, M.M., Easa, S.M., Batayneh, R.R.: Disaggregate gap acceptance model for unsignalized T-intersections. J. Transp. Eng. **123**(1), 36–42 (1997)
10. Rakha, H., Sadekb, S., Zohdyb, I.: Modeling differences in driver left-turn gap acceptance behavior using bayesian and bootstrap approaches. Procedia Soc. Behav. Sci. **16**, 739–750 (2011)
11. Adebisi, O., Sama, G.N.: Influence of stopped delay on driver gap acceptance behavior. Transp. Eng. **115**(3), 305–315 (1989)
12. Tian, Z.Z., et al.: Further investigation on critical gap and follow-up time. In: 4th International Symposium on Highway Capacity. Transportation Research Board, pp. 377–408 (2000). https://trid.trb.org/view/657360
13. Raff, M.S.: A volume warrant for urban stop signs. In: The Eno Foundation for Highway Traffic Control: Bureau of Highway Traffic Control, Yale University, Yale (1950)
14. Tanackov, I., Deretić, N., Bogdanović, V., Ruškić, N., Jović, S.: Safety time in critical gap of left turn manoeuvre from priority approach at TWSC unsignalized intersections. Physica A **505**, 1196–1211 (2018)
15. Pawar, D.S., Patil, G.R.: Analyzing variations in spatial critical gaps at two-way stop controlled intersections using parametric and non-parametric techniques. J. Traffic Transp. Eng. (Engl. Ed.) **8**(1), 129–138 (2021)
16. Witt, M., Kompaß, K., Wang, L., Kates, R., Mai, M., Prokop, G.: Driver profiling - data-based identification of driver behaviour dimensions and affecting driver characteristics for multi-agent traffic simulation. Transp. Res. Part F **64**, 361–376 (2019)
17. Patil, G., Sangole, J.: Behavior of two-wheelers at limited priority uncontrolled T-intersections. Int. Assoc. Traffic Safety Sci. Res. **40**, 7–18 (2016)
18. Barchański, A.: Analysis of critical gap times and follow-up times at selected, median, uncontrolled T-intersections differentiated by the nature of the surrounding. In: Macioszek, E., Sierpiński, G. (eds.) TSTP 2019. AISC, vol. 1083, pp. 242–256. Springer, Cham (2020). https://doi.org/10.1007/978-3-030-34069-8_19
19. Dissanayake, S., Lu, J.J., Yi, P.: Driver age differences in day and night gap acceptance capabilities. Int. Assoc. Traffic Safety Sci. Res. **26**(1), 71–79 (2002)
20. DeJoy, D.M.: An examination of gender differences in traffic accident risk perception. Accid. Anal. Prev. **24**(3), 237–246 (1992)
21. Yan, X., Radwan, E., Guo, D.: Effects of major-road vehicle speed and driver age and gender on left-turn gap acceptance. Accid. Anal. Prev. **39**, 843–852 (2007)
22. Macioszek, E.: Empirical analysis of gap acceptance parameters at roundabouts located in Tokyo (Japan) and the Tokyo surroundings. In: Macioszek, E., Kang, N., Sierpiński, G. (eds.) TSTP 2019. LNITI. Springer, Cham (2020). https://doi.org/10.1007/978-3-030-39109-6_1
23. Macioszek, E.: Analysis of significance of differences between psychotechnical parameters for drivers at the entries to one-lane and turbo roundabouts in Poland. In: Sierpiński, G. (ed.) Intelligent Transport Systems and Travel Behaviour. AISC, vol. 505, pp. 149–161. Springer, Cham (2017). https://doi.org/10.1007/978-3-319-43991-4_13
24. Dutta, M., Ahmed, M.A.: Gap acceptance behavior of drivers at uncontrolled T-intersections under mixed traffic conditions. Mod. Transp. **26**(2), 119–132 (2017). https://doi.org/10.1007/s40534-017-0151-9
25. Zohdy, I., Sadek, S., Rakha, H.: Empirical analysis of effects of wait time and rain intensity on driver left-turn gap acceptance behavior. Transp. Res. J. Transp. Res. Board **2173**, 1–10 (2010)

26. Serag, M.S.: Gap-acceptance behaviour at uncontrolled intersections in developing countries. Malays. J. Civ. Eng. **27**, 80–93 (2015)
27. Kaysi, I.A., Abbany, A.S.: Modeling aggressive driver behavior at unsignalized intersections. Accid. Anal. Prev. **39**, 671–678 (2007)
28. Patil, G.R., Pawar, D.S.: Temporal and spatial gap acceptance for minor road at uncontrolled intersections in India. Transp. Res. **2461**, 129–136 (2014)
29. Deo, P., Ruskin, H.J.: Urban signalised intersections: impact of vehicle heterogeneity and driver type on cross-traffic manoeuvres. Physica A **405**, 140–150 (2014)
30. Aakre, E., Aakre, A.: Modeling cooperation in unsignalized intersections. Procedia Comput. Sci. **109C**, 875–880 (2017)
31. Hobbs, F., Richardson, B.: Traffic Engineering, vol. 2. Pergamon Press, London (1967)
32. Gaca, S., Suchorzewski, W., Tracz, M.: Inżynieria ruchu drogowego. Teoria i praktyka. Wydawnictwo Komunikacji i Łączności, Warszawa (2011)
33. Bunden, W.: Introduction to Traffic Science. Printerhall Ltd., London (1967)
34. Tamhane, A.C.: Statistical Analysis of Designed Experiments. Theory and Applications. Wiley, New Jersey (2009)
35. Verbeek, M.: A Guide to Modern Econometrics, 2nd edn. Wiley, Chichester (2004)
36. Wikivoyage. https://pl.wikivoyage.org/
37. Mapa Targeo. https://mapa.targeo.pl/
38. Green, W.H.: Econometric Analysis. 7th edn. Prentice Hall, Hoboken (2012)
39. Aczel, A.D., Sounderpandian, J.: Statystyka w Zarządzaniu (In polish: Complete Business Statistics). PWN, Warszawa (2021)

Development of Ecological Public Transport in Poland on the Example of Selected Cities

Renata Krajewska, Ewa Ferensztajn-Galardos(✉), and Zbigniew Łukasik

Faculty of Transport, Electrical Engineering and Computer Science, Kazimierz Pulaski University of Technology and Humanities in Radom, Radom, Poland
{r.krajewska, e.ferensztajn, z.lukasik}@uthrad.pl

Abstract. Transport plays a very important role in the economy. The efficient movement of both people and freight largely influences the attractiveness of countries, regions and cities. At the same time transport imposes costs on society: greenhouse gas and pollutant emissions, noise, road accidents and congestion. Currently, transport emissions account for around a quarter of total greenhouse gas emissions in the EU. The changes taking place in the market, the growing environmental awareness and requirements of service users and the increasing restrictions on environmental protection are forcing the need to change the current model of development of urban transport systems to a more ecological one. The aim of this article is to analyse the conditions forcing changes in the functioning of public transport systems and to indicate possibilities of their financing. In the first part of the article, EU and national documents are presented, which define the actions to be taken within the framework of sustainable urban mobility, taking into account the aspect of environmental protection, competitiveness and satisfaction of social needs. The second part presents the possibilities of supporting pro-ecological activities from various sources in terms of innovativeness and competitiveness of transport systems.

Keywords: Ecological transport · Public transport · Low-emission transport modes · Financing sources

1 Introduction

Pro-ecological activities are currently conducted by an increasing number of business entities, because by focusing on various types of environmentally friendly solutions, companies providing transport services can achieve many economic benefits, and above all, maintain their position on the market, thus competing with other entities operating in the same industry. In addition, changes taking place on the market, growing awareness and requirements of customers and increasing restrictions on environmental protection enforce the need to change the current socio-economic development model to a more sustainable one. Actions to improve the region's transport accessibility are determined by a number of factors. Among them, ecology, which shapes the development and competitiveness of the region's economy, is an important priority.

E. Macioszek and G. Sierpiński (Eds.): TSTP 2021, LNITI, pp. 167–185, 2022.
https://doi.org/10.1007/978-3-030-93370-8_12

Sustainable development, in its ecological aspect (eco-development), is perceived as a process of reducing pressure on the environment and improving its condition through greening economic processes and implementing integrated environmental protection systems. Among the transport ecologisation factors we may enumerate: technical modernisation of means of transport, road and motorway networks, as well as development of environmentally-friendly branches of transport, e.g. railway transport as an alternative for car transport.

Mitigating the negative impact that transport generates on the environment is becoming an increasingly urgent issue. The EU recently adopted a climate and energy package which sets a target to reduce the level of greenhouse gas emissions in the EU. Transport plays a key role in achieving this objective and some existing trends will need to be reversed.

A modern transport system must be sustainable from the point of view of four aspects: economic, social, spatial, environmental. Efficient transport systems are essential for improving the welfare of society, as they have a significant impact not only on the rate of economic growth, but also on social development and environmental protection. A sustainable transport system:

- enables the basic need for individuals and societies to have access to it (the transport system) to be met, safely and consistently with the needs of human health and ecosystems, and responds to the requirements of equity values within and across generations,
- is affordable, operates efficiently, offers a choice of transport modes and supports a thriving economy,
- it reduces emissions and waste, taking into account the capacity of the ecosystem to absorb them, minimises the use of non-renewable resources, reduces the consumption of renewable resources to sustainable levels, recycles and reuses components thereof and minimises land use, and reduces noise.

The aim of this paper is to analyse the conditions forcing changes in the functioning of public transport systems and to indicate the possibilities of financing them. In the first part of the article, EU and national documents are presented which define the actions to be taken within the framework of sustainable urban mobility, taking into account the aspect of environmental protection, competitiveness and satisfaction of social needs. The second part presents the possibilities of supporting environmentally friendly activities from various sources in terms of innovativeness and competitiveness of transport systems. The third chapter gives examples of the implementation of low- and zero-emission transport modes in urban transport.

2 Strategies for the Development of Clean Public Transport in the EU and Poland

2.1 EU Documents Setting Out Actions for Sustainable Urban Transport

The deployment of sustainable and innovative transport modes plays an important role for the EU's climate and energy goals. With the increasing mobility of modern societies, EU policy focuses on supporting transport systems to address their main problems, which include [1]:

- congestion: it paralyses both road and air traffic,
- sustainability: most energy needs in transport are still met by oil, which is economically and environmentally unsustainable,
- air quality: The EU is committed to reduce emissions from trans-ports by 60% of 1990 levels by 2050 and to further reduce pollution emitted by vehicles,
- infrastructure: the quality of transport infrastructure varies between member states
- competition: Europe's transport sector faces increasing competition from fast-growing global transport markets.

One of the basic EU documents setting out actions for sustainable urban mobility is the Communication from the Commission to the European Parliament, the Council, the European Economic and Social Committee and the Committee of the Regions - Action Plan on Urban Mobility COM(2009) 490 final, 30.9.2009). The Communication proposes short and medium-term actions to ensure that urban transport in the European Union (EU) is environmentally sustainable, competitive and meets social needs. This Communication proposes the following solutions [2]:

- supporting integrated policies: urban transport management should be linked to other policies such as environment, urban planning and housing;
- taking citizens' interests into account: the public should be guaranteed: reliability, information, safety and accessibility to all modes of urban transport;
- cleaner urban transport: the EU will continue to support the development of new technologies for clean vehicles and alternative fuels, and the development of markets for these technologies and fuels;
- strengthening financing: given that most expenditure is covered by local, regional or national authorities, the European Commission will help them find EU funding;
- sharing experience and knowledge; the Commission will help all parties involved in public transport issues to benefit from existing experience and support the exchange of information;
- optimising urban mobility: The Commission will help local authorities to optimise the efficiency of their transport logistics and develop intelligent transport systems.

This document is linked to Directive 2009/33/EC of the European Parliament and of the Council of 20 June 2019 on the promotion of clean and energy-efficient road transport vehicles [3]. It aims to promote and stimulate the development of the market for clean and energy-efficient vehicles. It obliges public authorities and certain other public transport organisers to take the following factors into account (in relation to the whole life cycle of the vehicle) when purchasing vehicles: energy consumption, CO2 emissions and emissions of other pollutants.

The Directive applies to contracts for the purchase of road transport vehicles by contracting authorities and contracting entities and economic operators fulfilling public service obligations within the framework of a public service contract. These purchases concern clean and energy-efficient road transport vehicles. EU countries shall ensure that contracting authorities and entities and public service operators take into account energy and environmental impacts during the life cycle of the vehicle when purchasing road transport vehicles, namely: energy consumption; CO2 emissions; NOx, non-methane hydrocarbon (NMHC) and particulate matter emissions.

Several initiatives are currently underway to implement this directive. In this respect, we can mention:

- guidelines for green public procurement and a general technical report,
- a European Green Cars Initiative to promote the development of clean vehicles and sustainable mobility solutions,
- a European initiative on clean buses.

As far as the basic documents on the development of environmentally friendly transport in the EU are concerned, the non-legislative communication of the European Commission from 2011 stands out. COM (2011) 144 “White Paper - Roadmap to a Single European Transport Area - Towards a competitive and resource efficient transport system” [4]. The document is a kind of blueprint outlining 40 specific areas of action and listing 131 concrete initiatives for the next decade. It aims to create a competitive transport system that removes major bottlenecks and enables people and goods to move efficiently and safely throughout the European Union (EU). The Plan sets out the following objectives for policy action [4]:

- halve the use of conventionally fuelled cars in urban transport by 2030, phase them out in cities by 2050; achieve essentially CO2-free logistics in major urban centres by 2030,
- achieve 40% low-carbon fuel use in air transport by 2050,
- reduce CO2 emissions from maritime bunker fuels by 40% by 2050,
- shifting 30% of road freight transport over 300 km to other modes such as rail or waterborne transport by 2030, and shifting more than 50% of this type of transport by 2050,
- a three-fold increase in the existing high-speed rail network by 2030. By 2050, the majority of medium-distance passenger transport should go by rail,
- create a fully functional EU-wide TEN-T core network by 2030, integrating different transport modes,
- connecting major airports to the rail network, major seaports and the rail and inland waterway systems by 2050,
- introduction of traffic management systems, e.g. rail and road,
- establishment of a multimodal transport information, management and payment system by 2020,
- halving road traffic casualties by 2020 and achieving near-zero road transport fatalities by 2050,
- move towards full application of ‘user-pays’ principles (i.e. users of infrastructure bear the costs of its maintenance) and polluter pays (polluters bear the costs of clean-up),

The strategy aims to create a competitive transport system that increases mobility, removes major barriers in key areas and stimulates growth and jobs by 2050. It is complemented by the following initiatives [5]:

- Horizon 2020, which supports research and innovation in intelligent, clean and integrated transport,
- the Connecting Europe Facility, which provides funding for 10 international transport infrastructure projects,
- the Roadmap for moving to a low-carbon economy by 2020, which identifies transport as a key sector to achieve this goal,
- various energy efficiency measures to cut Europe's primary energy consumption by 20% by 2020.

Urban transport issues are directly addressed in the Communication from the Commission to the European Parliament, the Council, the European Economic and Social Committee and the Committee of the Regions "Towards joint efforts to achieve competitive and resource efficient mobility in cities" COM/2013/0913 final [6]. In this document, the Commission has focused on intensifying support for European cities in addressing urban mobility challenges, eliminating fragmented approaches and creating a single market for innovative urban mobility solutions by addressing issues such as common standards and specifications or joint procurement.

The most recent document presenting a strategy for sustainable and intelligent mobility is the Communication from the Commission to the European Parliament, the Council, the European Economic and Social Committee and the Committee of the Regions "Sustainable and Smart Mobility Strategy - putting European transport on track for the future" COM(2020)789 final [7]. It is an action plan of 82 initiatives that will guide work over the next four years. The strategy lays the foundations for a green and digital transformation and for making the EU transport system more resilient to future crises. A smart, competitive, safe, accessible and affordable transport system will cut emissions by 90% by 2050 [7].

By implementing this strategy it will be possible to create a more efficient and resilient transport system that aims to reduce emissions in line with the objectives of the European Green Deal. The "Sustainable and Smart Mobility Strategy" assumes that all modes of transport are to become more sustainable and that clean alternatives are to be widely available. Incentives must be put in place to implement the transformation. It defines specific milestones for developing the European transport system towards a smart and sustainable future [7]:

Phase I, which runs until 2030, assumes that at least 30 million zero-emission cars will be on Europe's roads, 100 European cities will be climate-neutral, rail express services will double across Europe, planned collective journeys of less than 500 km should be carbon-neutral, automated transport will be introduced on a large scale and zero-emission seagoing vessels will be market-ready,

Phase II - by 2035 concerns the market introduction of large zero-emission aircrafts;

Phase III assumes that by 2050 almost all cars, vans, buses and new trucks will be zero-emission, that rail freight traffic will have doubled and that the multimodal Trans-European Transport Network (TEN-T) for sustainable and intelligent transport with high-speed connections will be fully operational.

To achieve its objectives, the strategy identified a total of 82 initiatives in 10 key action areas ("flagship initiatives"), which can be referred to:

- sustainable transport development including the dissemination of zero-emission vehicles, renewable and low-emission fuels, zero-emission airports and ports, sustainable inter- and intra-urban mobility, greening of freight transport, and greenhouse gas pricing,
- intelligent mobility through seamless, safe and efficient connections. Innovation and digitalisation will shape the way passengers and goods travel in the future if the right conditions are created,
- a resilient mobility or a more resilient Single European Transport Area for an inclusive network. Trans-port is one of the sectors most affected by the COVID-19 pandemic and many companies in this sector are experiencing huge operational and financial difficulties.

The European Union's achievement of its climate goals necessitates a clear downward trend in CO2 emissions from the transport sector. The strategy adopted will lead to a change in the way people and goods are transported in Europe and will make it easier to combine different modes of transport within the same journey.

2.2 Green Public Transport Development Strategies in Polish Documents

The basic strategic document on transport in Poland is the Transport Development Strategy to 2030 [8] - SRT, adopted by the Council of Ministers on 24 September 2019. It is a key planning document for the development of the transport sector in the medium-term perspective and an integral element of the coherent management system of national strategic documents. The SRT serves the implementation of the objectives set out in national documents of a higher order, including primarily the Long-term National Development Strategy 2030 [9] and the National Development Strategy 2020 [10].

The SRT relates to all transport sectors: road, rail, air, sea, inland waterways, urban and intermodal transport. The main objective of the national transport policy is to increase transport accessibility and to improve the safety of road users and the efficiency of the transport sector by creating a coherent, sustainable, innovative and user-friendly transport system in the local, national, European and global dimensions. The SRT aims at reducing the negative impact of transport on the environment and climate and improving its energy efficiency by implementing modern technological solutions and shaping new patterns of mobility. An important element of the SRT is the development of public transport. The document also foresees the need to take into account the observed and forecasted climate change in the transport policy and to undertake adaptation measures in this respect.

In order to achieve the objectives of the environmental policy, the following directions of the SRT intervention are most important [11]:

- building an integrated, interconnected transport network serving a competitive economy,
- improving the way the transport system is organised and managed,
- changes in individual and collective mobility,
- reducing the negative impact of transport on the environment.

The strategy includes provisions on a review of the necessary measures (legal, organisational and investment) to improve the functioning of public transport, the introduction of low-emission vehicles, as well as urban zones with traffic ban. By 2030, the government assumes a gradual replacement of the rolling stock used to provide public transport services with clean, low-emission vehicles adapted to the needs of the elderly and people with disabilities. It is also planned to gradually include public transport in the "Common Ticket" project, the aim of which is to create, by 2025, the possibility for a traveller to purchase one ticket for the entire railway journey (all trains connecting the point of departure with the point of arrival), regardless of the sales channel and carrier. This will increase interest in travelling by rail, thus indirectly supporting climate neutrality and reducing emissions from transport [11].

The SRT document includes a number of measurable performance indicators, which include limiting the growth of annual final energy consumption by the transport sector to a maximum of + 15% points (25.8 Mtoe) by 2030 compared to the base year 2017. There is also a target to be achieved in 2030 - a minimum of 220 public transport trips per inhabitant in urban areas (compared to 171 trips in 2016). A number of investment measures are to serve the implementation of environmental goals, among which it is indicated, inter alia, to modernise the rolling stock of all modes of transport in order to bring it up to the state corresponding to EU and national standards, as well as to improve its energy efficiency and meet environmental protection requirements. This is to mean the gradual dissemination of low-emission and energy-efficient vehicles, including urban ones, using alternative fuels and propulsion (in particular, electric and CNG and LNG-powered), together with the launch of a network of stations for charging or replacing electric batteries and refuelling with natural gas and hydrogen. The government would like to see the share of alternative fuel buses in the total number of buses serving urban transport in Poland increase from 4.2% in 2017 to as much as 16% in 2030 [11].

Another of the integrated strategies, which addresses issues related to low- or zero-emission transport, is the "National Environmental Policy 2030 - a strategy for development in the area of environment and water management" - PEP2030, adopted by the Council of Ministers on 16 July [12]. It is the most important strategic document in this area. The role of PEP2030 is to ensure the ecological security of Poland and a high quality of life for all inhabitants. Under the Specific Objective: Environment and health. Improve environmental quality and ecological safety (I) 8.2 Direction of intervention: Elimination of air pollution emission sources or significant reduction of their impact (I.2). In the area of transport, it provides for the following tasks [13]:

- creation and improvement of a legal framework conducive to the implementation of effective measures to improve air quality,
- dissemination of financial mechanisms for improving air quality,
- engaging the public in action to improve air quality by increasing public awareness and creating sustainable platforms for dialogue with social organisations,
- development and dissemination of technologies conducive to improving air quality,
- development of mechanisms to control low emission sources, favouring improvements in air quality.

Another national document supporting sustainable transport development is the National Urban Policy 2023 [14], adopted by Resolution No. 198 of the Council of Ministers on 20 October 2015. It refers to all aspects of the functioning of cities, including transport issues. The most important indications concern road investments, which should focus on completing the construction of the basic transport system of the city and in its functional area, especially to the extent that will enable the elimination of the need for transit and truck access to industrial districts through the centre. Another direction of road investments should be the modernisation of the existing system in order to ensure its consistency with the priorities of the city's transport policy in the field of traffic safety, ensuring priorities for public transport and cycling. Solutions involving changes in the manner of shaping street space (or traffic organisation in such streets), especially in inner-city areas, are also indicated as contributing to a change in transport behaviour in favour of limiting individual car traffic. One of the ways of such shaping of space is the so-called speed zoning with the use of engineering and organisational solutions.

In October 2020, an update to the National Urban Policy was proposed to adjust it to the latest development priorities in Poland and international trends in urban development. The update of the National Urban Policy will change the perspective of its operation from 2023 to 2030.

Over the five years since the adoption of the NPM 2023, certain trends in urban policy have strengthened, which should be reflected in the NPM 2030. In particular, issues related to air quality, climate change, Smart Cities or strengthening cities' resilience to socio-economic and environmental crises should be deepened to a greater extent than before.

The problem of low use of alternative fuels (except LPG) in transport is attempted to be solved in our country by the adoption by the Polish Parliament of the Act of 11 January 2018 on electro mobility and alternative fuels [15]. This regulation is the result of the implementation of changes proposed in the "National Policy Framework for the Development of Alternative Fuel Infrastructure" adopted by the Council of Ministers on 29 March 2017 [16]. The Act sets out the conditions for the development and rules for the deployment of alternative fuel infrastructure in transport, the rules for the provision of electric vehicle charging and refuelling services for natural gas vehicles. It also imposes information obligations and introduces the obligation to use zero-emission vehicles by companies providing public services. The document also creates rules for the operation of clean transport zones.

One of the obligations concerning public transport organisers and operators provided for in the Act is the requirement to contract the provision of public transport services only to entities where the share of zero-emission buses in the fleet of vehicles used in this transport will amount to at least respectively [15]:

- 5% - from 1 January 2021,
- 10% as from 1 January 2023,
- 20% as from January 1, 2025,
- 30% as from 1 January 2028.

The Act on Electro mobility and Alternative Fuels transposes into the Polish legal system Directive 2014/94/EU of the European Parliament and of the Council of 22

October 2014 on the development of alternative fuel infrastructure (Official Journal of the EU of 28 October 2014, item L 307/1).

European Union and national regulations on transport issues occupy an important place. Issues related to the sustainable development of urban transport and the reduction of transport emissions are particularly emphasised. Analysing the obligations to reduce emissions of air pollutants and taking into account the aim to achieve EU standards, one should expect an increase in innovation in the Polish economy, especially in the area related to electromobility both in the individual and collective dimension.

3 Sources of Finance for Clean Urban Transport Measures

The basic problem related to the functioning of ecological public transport in Poland is often the lack of financial resources enabling the development of pro-ecological activities. Therefore, funds available within particular operational programmes are an important source of co-financing for public transport systems. Sources of funding are money that cover the costs of business or initiatives. From the point of view of origin, internal and external sources of financing can be distinguished for activities carried out by transport enterprises, including environmental activities.

Examples of internal sources of financing for public transport enterprises may include net profit from operations, depreciation, and sale of redundant assets or other free resources. External financing, on the other hand, requires the involvement of third-party capital. These may be bank loans, leasing, and capital of new or existing owners, subsidies. The characteristic feature of this category of instruments is their repayable and remunerative nature and the external form of origin. However, external financing does not always take the form of a repayable instrument. A good example here may be public transport, which is subsidised from local government budgets and these are non-refundable funds.

The most important external sources of funding for investments in green urban rolling stock in Poland include:

- EU funds (Operational Programmes Infrastructure and Environment-POI&E, Development of Eastern Poland, Regional Operational Programmes),
- national funds from the National Fund for Environmental Protection and Water Management, National Green Investment Scheme,
- domestic and foreign loans and borrowings.

Funds for financing pro-ecological activities can be obtained from various sources, in various forms - as financial instruments in accordance with the principle of rationality, including the possibility of obtaining, as well as the level of necessary funds and sources of financing - the size of co-financing are dependent on the size of the enterprise, its organisational and legal form, the situation in the external environment of the entity - its position on the market, but above all on the implemented undertaking. In the financial perspective 2021–2027, Polish cities will be able to benefit from the nationwide - generally available operational programmes.

European funds for the years 2021–2027 amount to EUR 72.2 billion from the cohesion policy and EUR 3.8 billion from the Equitable Transition Fund. In total, this amounts to around €76 billion. The funds will be used to make investments in innovation, entrepreneurship, digitalisation, infrastructure, environment, energy, education and social affairs [17].

The Infrastructure and Environment Operational Programme is the largest programme financed by European Funds not only in Poland, but also in the European Union. It is a national programme supporting low-carbon economy, environmental protection, climate change mitigation and adaptation, transport and energy security as well as investments in health care and cultural heritage. The main objective of the Infrastructure and Environment Operational Programme is the development of technical infrastructure while protecting and improving the environment. This programme may finance not only the purchase of environmentally friendly means of transport, but also such initiatives as the modernisation and construction of lines and purchase of tramway rolling stock. These vehicles do not emit pollutants along the transport corridor, so they do not lower the quality of air in the city. In addition, modern trams have lower electricity consumption and noise levels and are equipped with recuperation systems allowing energy recovery during braking. What is important from an ecological point of view is that it can come from renewable sources, and thus allows independence from dependence on fossil fuels. The purchase of 400 trams serving the most important urban centres in Poland would not have been possible without approximately 48% co-financing from OPI&E [18].

Thanks to the balance between investment measures in infrastructure and support targeted at selected areas of the economy, the programme will effectively implement the assumptions of the Europe 2020 strategy, to which its main objective is linked - support for a resource-efficient and environmentally friendly economy, and support for territorial and social cohesion. Areas of support and types of projects possible to be implemented under the Infrastructure and Environment programme in the field of transport include:

- development of the TEN-T road network and multimodal transport development of road infra-structure in the TEN-T network, including: improvement of road traffic safety; improvement of air traffic safety and intermodal, maritime and inland waterway transport,
- road infrastructure for cities, including improvement of accessibility to cities and capacity of road infrastructure (development of road infrastructure in cities and exit routes from cities, construction of ring roads),
- development of railway transport in Poland, including the development of railways in TEN-T, outside the network, and urban railways,
- development of low-emission collective transport in cities, including infrastructure and rolling stock for public collective transport in cities and their functional areas.

The Operational Programme Development of Eastern Poland (OPDP) is aimed at levelling social and economic disproportions between the voivodships covered by the project and the remaining part of the EU. Under this programme measures are being implemented to reduce the adverse impact of the urban transport system on the environment. This financial instrument, which serves to implement the EU's cohesion policy, is used, among other things, for the purchase of clean urban vehicles. An example of this is the financing of modern trans-port infrastructure (Modern Transport Infrastructure) as part of measures aimed at subsidising sustainable urban transport. This programme provides funding for complex projects to create new or expand existing green integrated urban transport networks. It covers such tasks as:

- construction/reconstruction of bus, trolleybus and tram networks, together with the purchase of low-emission rolling stock,
- construction/reconstruction of necessary infrastructure for urban transport, including intermodal transfer stations,
- implementation of new, extension or modernisation of existing telematics systems for the needs of urban transport.

In total, 14 urban transport projects are implemented under the Eastern Poland Programme, in all voivodship towns of Eastern Poland: Lublin, Olsztyn, Kielce, Rzeszow and Bialystok. The projects will result in friendly, ecological public transport, well serving residents, entrepreneurs and visitors. Among other things, over 12 km of tram lines will be constructed as part of the investment, 314 units of rolling stock will be purchased and 14 components of intelligent transport systems (e.g. passenger information system) will be installed.

In addition, green public transport in Poland can be financed from the National Fund for Environmental Protection and Water Management under one of the programmes implemented in 2021- Zero emission transport [19]:

- Green Investment Scheme (GIS) - Kangaroo - Safe and ecological way to school,
- My Electric,
- Green public transport - Phase I,
- Support for electric vehicle charging and hydrogen refuelling infrastructure.

The beneficiaries of the programme implemented by the National Fund for Environmental Protection and Water Management may be organisers and operators of public mass transport, including local governments. Co-financing may also be allocated to the purchase of charging stations and staff training. Financial support is provided for two types of activities. - The first is the purchase of new zero-emission rolling stock, and the second is the construction of appropriate associated infrastructure. The Fund's money can be used to purchase or lease new electric buses powered solely by electricity accumulated by connecting to an external power source, together with training for drivers and mechanics in the operation of zero-emission vehicles. The fund will also be used to purchase trolleybuses and electric buses, which use electricity generated from hydrogen for propulsion.

The objective of the programme "green public transport" is to avoid emission of air pollutants by co-financing undertakings reducing the use of emission fuels in transport. The programme provides for the possibility of co-financing projects aimed at reducing the use of emission fuels in public collective transport [20]:

Concerning vehicles consisting in

- the acquisition/lease of new electric buses using only the electric energy accumulated by connecting to an external power source for propulsion, together with training of drivers/mechanics in the field of operation of zero-emission vehicles,
- purchasing/leasing new trolleybuses, i.e. buses adapted to be supplied with electric energy from the traction network and equipped with an additional propulsion system, thanks to which they will be able to cover a route without electric traction (e.g. traction batteries or hydrogen fuel cell) along with training for drivers/mechanics in the field of operation of zero-emission vehicles,
- purchase/leasing of new electric buses using exclusively electricity generated from hydrogen in the installed fuel cells for propulsion together with training of drivers/mechanics in handling of emission-free vehicles.

Modernisation and/or construction of infrastructure enabling the operation and proper use of the purchased/leased vehicles, including in particular hydrogen charging or refuelling points, together with the accompanying infrastructure necessary for their operation, or a traction network - the infrastructure will be used exclusively for the operation of public transport.

In financing projects related to ecology, environment protection or air quality improvement, the banking sector and green bonds may be an alternative to EU funds. Green bonds are no different from classic bonds, except for the purpose of the issue, which can be low-emission urban transport or a cycle path, among others. Green bonds are beneficial both from the perspective of the issuing cities and investors. Cities use them to promote their key green investments, while investors - banks or investment funds - look for such products, as corporate policy dictates that they should invest a certain percentage of their capital in investments related to ecology and environmental protection.

The issue of green bonds starts with the city and the bank preparing a catalogue of green investments planned within the next few years. The city then approves which of these projects it would like to implement. The bank prepares a credit decision which gives the final answer as to whether a given project can be financed in such an amount and over such a period of time. This is where an external institution comes in - usually a European NGO, which grants a 'green' certificate, i.e. it confirms that the given investment is indeed related to ecology and confirms the purpose of the bond issue. Then we issue bonds, find investors, or banks themselves purchase such bonds for their own portfolios, and the city receives funds to finance these projects.

Polish cities will become greener. Reducing emission-intensive transport and cooperating with renewable energy will be investments preferred by cities in the perspective until 2030. However, due to the fact that the next new EU perspective will not be as generous for local governments, alternative sources of financing such projects should be sought. The capital necessary to implement such measures may be obtained by issuing green bonds - securities which finance pro-ecological investments, e.g. wind power plants or environmentally-friendly transport. Polish cities are increasingly focusing on ecology and solutions aimed at reducing carbon dioxide emissions, increasing the share of low-emission public transport and making the inhabitants' lives more pleasant. In the financial perspective until 2030, the majority of investments will concern low-emission transport, including the purchase of electric and hybrid buses.

According to the forecasts included in the report "Alternative fuels in municipal transport" of the PSPA, within the next 10 years about 3.5 thousand buses powered by alternative fuels will appear on the streets of Polish cities.

Compared to European cities, there is still a lot to be done in Poland in terms of pro-ecological activities, but a lot has already been done, as the previous EU perspective made it possible to co-finance projects in the field of low-emission transport, improvement of the quality of life of inhabitants and energy efficiency.

4 Examples of Implementation of Low-and Zero-Emission Means of Transport in Selected Polish Cities

For years, both the recommendations and other documents of the European Commission have emphasised the importance of a sustainable transport policy as the solution that best counteracts the negative environmental and social effects. Moreover, many researchers dealing with transport on a daily basis draw attention to the fact that the idea of adjusting urban transport supply to the needs of motorised traffic is doomed to failure. The prevailing view in this environment is that the demand for this mode of transport should be adapted to the current supply and to environmental and social requirements. This can be achieved by implementing sustainable solutions for urban transport in terms of:

- means of transport technology,
- collective transport,
- individual transport,
- freight transport,
- traffic organisation and management.

Sustainable, safe, environmentally friendly and energy efficient urban transport of the future is based on appropriate means of transport. Since in the future more than 70% of the world's population will live in urban areas, the main focus in the coming decades should be on reducing emissions of toxins contained in exhaust gases, reducing CO_2 content, increasing energy efficiency of transport and reducing noise levels in urban transport. Advances in transport technology can be made in vehicle, engine and fuel technology [21].

In public transport, buses are still the most popular and widely used means of transport. They are present practically everywhere where public transport operates. According to the Central Statistical Office, almost 12,169 public transport buses were in operation in Poland at the end of 2019. The largest urban bus operators in Poland are: MZA in Warsaw (1,425 vehicles), MPK in Krakow (566), MPK in Lodz (408), GAiT (320), MPK in Poznan (316) and MPK in Wroclaw (296) [22].

The evolution of "green" technologies in transport is particularly visible in the case of public transport. An example are buses, where gaseous fuels (CNG, LNG, bio methane) are becoming an energy source in addition to traditional fuels - engines fuelled by natural gas in the form of CNG or LNG emit significantly less toxic compounds than their equivalents fuelled by diesel or petrol. The basic fuel supplying Polish buses is still diesel oil. There are relatively few vehicles powered by other fuels -

at the end of 2017, the fleet of municipal carriers included 64 CNG buses, 40 LNG buses, 20 biofuel buses, and 104 hybrid vehicles. Increasingly, the fleet of conventionally powered vehicles is supplemented by all-electric buses, of which there were 52 in 2017, but their share is increasing year on year. Electric buses are emission-free at the point of use. Although most of the electricity to charge them is produced in coal-fired power plants in Poland, electric vehicles are more environmentally friendly than combustion vehicles meeting the most stringent emission standards. They are also quiet, which eliminates another problem in cities - noise, generated in about 90% by road transport. Table 1 presents the advantages and disadvantages of different types of alternative-powered buses.

Table 1. Advantages and disadvantages of different types of alternative-powered buses [23].

Powered	Advantage	Disadvantage
Gas	Low fuel costs	Necessary access to gas network
Electric	Low electricity costs Emission-free Quiet EU co-financing	High purchase costs (up to 2.5 times higher than for diesel) High battery replacement costs Necessary infrastructure Need to adapt the bus fleet model to the needs
Hybrid	Lower fuel costs Partly emission-free	High purchase costs (up to 2.5 times higher than for diesel) Technical complexity Savings depending on the right track
Hydrogen	Emission-free Quiet Longer range than an electro bus	High purchase costs (up to 2.5 times higher than for diesel) Lack of required infrastructure in Poland

Electric buses, although currently more expensive than normal diesel buses, allow relatively cheap introduction of electric transport in cities: they do not require very expensive investments in infrastructure development. Electro buses require only spot investments in recharging infrastructure, but the variety of batteries and methods of charging allows to shape this infrastructure according to local needs. With suitably selected batteries and route parameters, operators can limit themselves to building, for example, chargers at the depot. Other advantages of electro buses include significantly lower noise level emitted by the vehicles, lower vibrations, smoother ride and more efficient and cheaper operation. Experts also point out that although the Polish energy sector is still based on coal, the spread of electric buses will significantly reduce its consumption.

The vehicles with an alternative drive to the combustion engine are used to the greatest extent by the Municipal Bus Company in Warsaw. In total, they use 35 buses powered by liquid gas, 4 hybrid and 31 electric buses. In the capital city there are also affiliate carriers providing services commissioned by the Public Transport Authority. Arriva owns, among others, 5 hybrids, Mobilis - 61 hybrids, and PKS Grodzisk

Mazowiecki - 5 buses running on biofuels [24]. Moreover, significant - as for Polish conditions - number of alternative-propulsion vehicles is also available in MPK Krakow (5 CNG, 12 hybrid, 26 electric buses), MPK Czestochowa (2 CNG and 39 hybrid), MZK Tomaso Mazowiecki (25 hybrid) and MZK Samos (26 CNG vehicles).

The number of means of transport of selected bus operators in Poland in 2018 including buses with alternative propulsion is shown in Table 2.

Table 2. Means of transport of selected bus operators in Poland in 2018, including buses with alternative propulsion [23].

Urban bus operators in Poland	Total number of buses	CNG buses	Hybrid buses	Electric buses	Biofuel buses	Average age of buses
MZA in Warsaw	1425	35	70	31	5	8
MPK in Krakow	566	5	12	28	0	6
MPK in Lodz	408	0	0	0	0	8
GAiT in Gdansk	320	0	0	0	0	9
PKM in Katowice	258	0	0	11	0	7
MPK in Lublin	209	0	0	2	0	10
MPK in Czestochowa	206	2	40	0	0	8
MZK in Bydgoszcz	217	0	0	0	0	6
MPK in Radom	128	42	0	0	0	12

The authorities of individual cities and, most often, the transport companies owned by them, are introducing more and more ecological solutions in transport. This is enforced by national and EU legislation, in particular by the Euro VI standard which must be met by conventional combustion vehicles that constitute a significant part of the bus fleet in Poland. For this reason it can be expected that in the coming years the importance of zero-emission buses - both electric vehicles and those using hydrogen cells - will grow.

A large part of the Polish bus fleet consists of modern ecological buses equipped with engines meeting the highest Euro 6 emission standard. However, many cities are implementing investments concerning the purchase of zero-emission or low-emission rolling stock co-financed from various available EU programmes. Examples are presented in Table 3.

A stable perspective of financial support allows not only cities to plan gradual replacement of their fleets, but also polish bus manufacturers to smoothly develop their order portfolios. Applications for co-financing in the first stage can be submitted on a continuous basis until mid-December 2021, but no longer than until the allocation funds are exhausted. In this year's first stage, public transport operators, including local government units, may obtain grants only for purchasing or leasing electric buses and trolleybuses - in both cases, the maximum share of co-financing in the purchase/leasing costs may amount to 80%. In the second call in 2022, the level of subsidies for electric buses and trolleybuses will decrease by 10% points, and there will be a possibility to obtain grants up to 90% for purchasing hydrogen-powered buses.

Table 3. Purchase of low- and zero-emission buses in selected cities between 2020 and 2022 co-financed by EU funds [25–37].

Bus operators	Planned purchase of transport vehicles	Source of funding and amount of co-financing
MZA in Warsaw	130 electric buses 70 CNG buses by Solaris 90 low-emission buses fuelled by liquefied natural gas	Operational Programme Infrastructure and Environment 2014–2020 - PLN 180 million
MPK in Krakow	50 electric buses	Operational Programme Infrastructure and Environment 2014–2020 - PLN 110 million
MPK in Lodz	17 electric buses	Łódzkie Regional Operational Programme - PLN 40 million
GAiT in Gdansk	26 electric buses 49 zero-emission buses	National Fund for Environmental Protection and Water Management within the framework of Green Public Transport - will finance about 80% of the expenses, i.e. PLN 240 million
PKM in Katowice	32 electric buses	The subsidy will come in half from the Regional Operational Programme of the Silesian Voivodship for 2014–2020 and in half from the GEPARD II programme implemented by the National Fund for Environmental Protection and Water Management - PLN 80 million
MPK in Lublin	12 electric buses	Eastern PolandOperational Programme 2014–2020 Priority axis II: Modern transport structure Measure 2.1 Sustainable urban transport - PLN 32 million
MPK in Czestochowa	15 electric buses	European Regional Development Fund - PLN 7.5 million
MZK in Bydgoszcz	30 electric buses	The National Fund for Environmental Protection and Water Management within the framework of Green Public Transport will finance about 60% of the expenses, i.e. about 37 million PLN
MPK in Radom	19 electric buses	Operational Programme Infrastructure and Environment 2014–2020 - PLN 19.16 million

In turn, in the third call in 2023, it will again be possible to obtain subsidies only for battery-powered buses and trolleybuses. In both cases, the level of subsidies will be reduced to 60%. Moreover, in each of the calls it will be possible to obtain subsidies covering up to 50% of the costs of purchase of charging infrastructure, but their amount cannot exceed PLN 3 million [38].

5 Conclusions

An important element of the pro-ecological policy implemented by the European Union is support for the development of enterprises, in particular small and medium-sized enterprises (SMEs) in terms of successive increasing their innovativeness and competitiveness in the context of introducing pro-ecological solutions. The introduction of effective pro-ecological solutions, including the purchase of ecological means of transport in cities, on the one hand, favours not only the development of enterprises providing transport services for passengers - public transport, but the entire economy, however, it is connected with incurring significant costs.

On the basis of the analysis made, it can be stated that the EU funds are an important source of financing environmentally-friendly activities of transport enterprises and thus enable their socio-economic development. Both the identification and implementation of the pro-ecological strategy of enterprises is an important condition for increasing the rationality of management, which, in accordance with the principle of sustainable development, should take into account economic efficiency, ecological rationality and social legitimacy, as it is an interdisciplinary development focused on economic issues - whose dimension is effectiveness, i.e. better use of available resources, including financial ones, ecological issues - whose overriding aim is to prevent or reduce the negative impact on the environment, and social issues focused on social responsibility for the effects of business activity.

So far, thanks to the support from EU funds and national programs such as Gepard or Kangaroo, local governments have contracted the purchase of about 690 electric buses, of which 430 are already registered. Undoubtedly, the challenge of electrification of urban public transport is enormous, as there are over 12,000 buses in Polish cities, and their average mileage is over 75,000 km. That is why the National Fund for Environmental Protection and Water Management is already working on the second phase of the Green Public Transport programme, to be financed from the National Recovery Plan.

References

1. Transport in EU. https://europa.eu/european-union/topics/transport_pl
2. COM (2009) 490: Communication from the Commission to the European Parliament, the Council, the European Economic and Social Committee and the Committee of the Regions - Action Plan on Urban Mobility, Brussels, 23 September 2009. https://eur-lex.europa.eu/LexUriServ/LexUriServ.do?uri=COM:2009:0490:FIN:EN:PDF

3. Directive (EU) 2019/1161 of the European Parliament and of the Council of 20 June 2019 amending Directive 2009/33/EC on the promotion of clean and energy-efficient road transport vehicles (Text with EEA relevance). https://eur-lex.europa.eu/eli/dir/2019/1161/oj
4. COM (2011) 144: WHITE PAPER Roadmap to a Single European Transport Area - Towards a competitive and resource efficient transport system, Brussels, 28 March 2011. https://eur-lex.europa.eu/LexUriServ/LexUriServ.do?uri=COM:2011:0144:FIN:EN:PDF
5. Roadmap to a single European transport area: towards a competitive and resource-efficient transport system. https://eur-lex.europa.eu/legal-content/PL/TXT/?uri=legissum%3Atr0054
6. COM (2013) 0913 final: Communication from the Commission to the European Parliament, the Council, The European Economic and Social Committee and the Committee of the Regions Together towards competitive and resource-efficient urban mobility. https://eur-lex.europa.eu/legal-content/EN/TXT/?uri=CELEX%3A52013DC0913
7. COM (2020) 789 final: Communication From The Commission To The European Parliament, The Council, The European Economic And Social Committee And The Committee Of The Regions sustainable And Smart Mobility Strategy-Putting European Transport On Track For The Future, Brussels, 9 December 2020. https://eur-lex.europa.eu/legal-content/EN/TXT/?uri=COM:2020:789:FIN
8. Resolution No. 105/2019 of the Council of Ministers of 24 September 2019 on the adoption of the Strategy for Sustainable Transport Development until 2030 (M.P. item 1054) (2019). http://isap.sejm.gov.pl/isap.nsf/download.xsp/WMP20190001054/O/M20191054.pdf
9. Resolution No. 16 of the Council of Ministers of 5 February 2013 on the adoption of the Long-term National Development Strategy. Poland 2030. The Third Wave of Modernity (M. P. item 121) (2013). http://isap.sejm.gov.pl/isap.nsf/download.xsp/WMP20130000121/O/M20130121.pdf
10. Resolution No. 157 of the Council of Ministers of 25 September 2012 on the adoption of the National Development Strategy 2020 (M.P. item 882) (2012). http://isap.sejm.gov.pl/isap.nsf/download.xsp/WMP20120000882/O/M20120882.pdf
11. Strategy for Sustainable Transport Development until 2030, Adopted with the Resolution No. 105/2009 of the Council of Ministers as of 24 September 2019 (M.P. Item 1054) (2019). https://www.gov.pl/web/ia/strategia-zrownowazonego-rozwoju-transportu-do-2030-r-srt2030
12. Resolution No. 67 of the Council of Ministers of 16 July 2019 on the adoption of the National Ecological Policy 2030 - development strategy in the field of environment and water management (M.P. item 794) (2019). http://isap.sejm.gov.pl/isap.nsf/download.xsp/WMP20190000794/O/M20190794.pdf
13. The 2030 National Ecological Policy – the Development Strategy in the Area of the Environment and Water Management, Ministry of Climate and Environment, Warsaw (2019). https://bip.mos.gov.pl/fileadmin/user_upload/bip/strategie_plany_programy/Polityka_Ekologiczna_Panstwa/Polityka_Ekologiczna_Panstwa_2030.pdf
14. National Urban Policy to 2023, Warsaw (2015). https://www.gov.pl
15. Law of 11 January 2018 on electromobility and alternative fuels (M.P. item 908) (2021). https://isap.sejm.gov.pl/isap.nsf/download.xsp/WDU20180000317/U/D20180317Lj.pdf
16. National Policy Framework for the Development of Alternative Fuel Infrastructure, Ministry of Energy, Warsaw (2017). https://www.gov.pl
17. European Funds 2021–2027. https://www.funduszeeuropejskie.gov.pl
18. European Funds - Partnership Agreement. https://www.funduszeeuropejskie.gov./plmedia/97650/umowa_partnerstwa_broszura_wersja_dostepna.pdf
19. National Fund for Environmental Protection and Water Management-Programmes. http://www.nfosigw.gov.pl/oferta-finansowania/srodki-krajowe/programy-2021/

20. National Fund for Environmental Protection and Water Management - Green Public Transport. http://nfosigw.gov.pl/oferta-finansowania/srodki-krajowe/programy-priorytetowe/zielony-transport-publiczny-faza-i/
21. Piernikarski, D.: Zrównoważony system transportowy: moda czy konieczność? Autobusy **11** (6), 1–12 (2010)
22. Transport. http://swaid.stat.gov.pl/TransportLacznosc_dashboards/Raporty_predefiniowane/RAP_DBD_TRANS_11.aspx
23. Poland on The Road to Electromobility: TOR Zespół Doradców Gospodarczych. Transport Publiczny, Warsaw (2018)
24. Statistical handbook, no I (322). Zarząd Transportu Miejskiego w Warszawie (2021). https://www.ztm.waw.pl/wp-content/uploads/2021/03/www-BIULETYN-STYCZE%C5%83-2021.pdf
25. Bydgoszcz - purchase of buses. https://metropoliabydgoska.pl/trwaja-przygotowania-do-zakupu-autobusow-niskoemisyjnych-maja-jezdzic-na-trzech-liniach/
26. Electric buses in Tri-City. https://moto.trojmiasto.pl/Autobusy-elektryczne-w-Trojmiescie-n150683.html
27. MPK Czestochowa. http://m.infobus.pl/mpk-czestochowa-wybiera-15-elektrobusow-autosana-_more_126389.html
28. MPK Krakow. http://www.pzptp.pl/aktualnosci/artykul/mpk-krakow-z-umowa-na-dostawe-50-przegubowych-elektrobusow
29. MPK Łódź. http://www.mpk.lodz.pl/showarticle.action?article=25500&17+autobus%C3%B3w+elektrycznych+Volvo+trafi+do+%C5%81odzi
30. MZA Warsaw. http://www.pzptp.pl/aktualnosci/artykul/90-gazowych-autosanow-w-mza-w-warszawie
31. Public transport in Gdansk. https://www.transport-publiczny.pl/wiadomosci/gdansk-czekamy-na-fundusz-niskoemisyjnego-transportu-66367.html
32. Public transport in Katowice. https://www.transport-publiczny.pl/mobile/jest-umowa-pkm-katowice-z-solarisem-na-dostawe-5-elektrobusow-65518.html
33. Public transport in Krakow. https://www.transport-publiczny.pl/wiadomosci/krakow-pierwsze-nowe-elektrobusy-solarisa-z-umowy-na-50-pojazdow-przekazane-68027.html
34. Public transport in Radom. https://www.transport-publiczny.pl/wiadomosci/radom-kupuje-kolejnych-dziewiec-elektrycznych-solarisow-jest-umowa-65658.html
35. Czestochowa - Purchase of buses. https://www.cire.pl/item,210961,1,0,0,0,0,0,czestochowa-z-unijna-dotacja-kupi-4-autobusy-elektryczne.html
36. Zero-emission buses in Warsaw. https://www.um.warszawa.pl/aktualnosci/130-zeroemisyjnych-autobus-w-dla-warszawy
37. ZTM Lublin. http://www.pzptp.pl/aktualnosci/artykul/ztm-w-lublinie-podpisal-umowe-z-solarisem
38. National Fund for Environmental Protection and Water Management. www.nfosigw.gov.pl

Assessment of External Economic Cooperation in the Aspect of Transport Development of the Greater Tumen Initiative

Vladimir Nekhoroshkov[1], Sergey Vakulenko[2], Peter Kurenkov[2], Evgeniy Nekhoroshkov[1], Grigoriy Deruzhinskiy[3(✉)], Alexander Ignatenko[3], Alyona Aroshidze[1], Alexey Astafiev[2], Irina Seryapova[4], and Irina Solskaya[5]

[1] Siberian Transport University, Novosibirsk, Russia
en@stu.ru
[2] Russian University of Transport (RUT–MIIT), Moscow, Russia
post-iuit@bk.ru
[3] Admiral F. F. Ushakov State Maritime University, Novorossiysk, Russia
evropa@bk.ru
[4] Samara State University of Railway Transport, Samara, Russia
[5] Irkutsk State Transport University, Irkutsk, Russia
solskaya_i@irgups.ru

Abstract. Economic cooperation among the countries of the Greater Tumen Initiative (GTI) is the basis for interaction in various fields, especially transport. It is obvious that transport infrastructure development and the border barriers influence reduction of are prerequisites for trade stimulation and developing other forms of regional integration. On the other hand, the transport corridors development, the borders control modernization, transshipment points, etc. - all these are dependent on their congestion with foreign trade flows. Consequently, the development and feasibility of transport projects implementation within Greater Tumen Initiative should count on the potential of foreign economic cooperation between member countries and the possible degree of interconnection of economies. This article presents the results of a study in the complementarity assessment of foreign economic cooperation within the framework of the Greater Tumen Initiative, its potential, and the foundations for understanding the realistic possibilities to implement transport projects that are aimed at servicing these commodity flows. In addition, an analysis of the current state of the transport infrastructure of cross-border cooperation is presented. Along with the assessment results, this study reflects the feasibility of cooperation in the transport sector within GTI and highlights priority areas to benefit Russia's interests. The results of this study contribute in understanding of the problems and prospects for the GTI transport projects development and implementation taking into account the importance of an unstable geopolitical, financial and economic background.

Keywords: Greater Tumen Initiative (GTI) · Transport sector · Foreign economic cooperation · Complementarity

E. Macioszek and G. Sierpiński (Eds.): TSTP 2021, LNITI, pp. 186–199, 2022.
https://doi.org/10.1007/978-3-030-93370-8_13

1 Introduction

The Greater Tumen Initiative (GTI) is an intergovernmental system of cooperation between four countries - China, Mongolia, the Republic of Korea, and the Russian Federation in Northeast Asia (the Democratic People's Republic of Korea was a member of the GTI until 2009). The governments of the GTI member countries have pledged to strengthen economic cooperation and promote sustainable development and growth in Northeast Asia and especially in the Greater Tumen region, including Chinese northeastern provinces - Jilin, Heilongjiang and Liaoning, and Inner Mongolia, along with eastern provinces of Mongolia, eastern port cities of the Republic of Korea, and Primorsky Krai of Russia.

Starting with the "Tumen" program and ending up with the GTI - the initiative develop the Tumen region has been existing for over 25 years and still does not lose its relevance for its member countries as well as for the entire Northeast Asia. It is interesting that most of the studies focused on the development of these countries do not consider the problems and prospects of their participation in the GTI.

For example, studies on the economic development of China as the GTI member country are devoted to a variety of aspects - from finding determinants and development problems to building specific models. Thus, the work of Lu et al. [1], He et al. [2] both are forecasting the impact of financial development and trade on the entire Chinese economy and building a model of its growth and at the same time devotes the impact of rising trade costs on the economic transformation, which is an important political practice for China's economic development. Among the studies for People's Republic of China there should be Fleisher and Zhang [3] review highlighted, that provides detailed data for the Chinese economy for over 30 years. The economic development of the Republic of Korea as another member of the GTI is studied, for example, from the standpoint of the comparative advantages manifested in international trade, which is discussed in the study by Sim and Yoo [4], shocks of foreign and domestic economic policy, the impact of which is studied by Cheng [5], integration processes that are studied by Lim and Breuer [6]. Chu et al. [7] study the development of another GTI member country - Mongolia. They have attempted to identify the determinants of growth for Mongolia, counting on the impact of its large neighbor - China. Among studies on the economic development of the entire Northeast Asia the study of Thorbecke [8] should be highlighted, that analysis the characteristics of East Asian products and services values chains and their dependence of various factors. The economic development of Russia is studied in several papers, among those Voskoboynikov [9] article is worth noting where the author tried to find sources of long-term economic growth. Mitsek [10] provides the similar study and presents Russian economy econometric model, that enables retrospective and predictive analysis of economy dynamics. In addition, such authors as Ono [11], who devoted his paper to establish the link between the financial crisis and economic growth in Russia, and Bayramov et al. [12], who studied the medium-term impact of economic sanctions on development not only in Russia, but also within the countries of the ex-USSR.

This review proves that the GTI member countries are one of the most interesting and relevant research objects, various economic issues of their growth and development hold leading places in the practice of research activities. The total gross domestic product (GDP) of the GTI member countries is more than 20% of the global value with the unconditional leadership of China and three out of four countries (China, the Republic of Korea, Russia) are among the world's key exporters with a combined share of more than 18%. Undoubtedly, these facts confirm the potential possibilities of the integration interaction of these countries for the transformation of the GTI into a full-fledged intergovernmental organization.

One of the key objectives of GTI is to promote the development of transport infrastructure. To achieve this goal the Transport Council was created, whose activities are aimed at improving transportation within the GTI and identifying the necessary steps for the development of transport corridors. The member countries agreed that the transport network should be considered as the main priority of cooperation, and it has to be focused on the key regional transport corridors development and construction to connect the main ports, railways, and highways within the Greater Tumen area.

The transport aspect in the context of the development of the GTI member countries is considered in Magazzino and Mele [13] paper, where the author attempted to find the relationship between transport infrastructure and the economic growth of China, and Foo et al. [14], who studied the impact of China's Belt and Road Initiative on international trade. It should be noted that the most common object of research on the transport subject of the GTI member countries is the Belt and Road megaproject, which, in accordance with its transit function, can also have a significant impact on the implementation of GTI transport projects. For example, it is quite realistic to connect Russia's initiatives to develop transport corridors, considering its interests in the GTI, with the concepts of the Silk Road Economic Belt and the 21st Century Maritime Silk Road. Lall and Lebrand [15] tried to understand the spatial effect of the Belt and Road Initiative and its differentiation, implications for the whole of Central Asia. Note that the relationship between transport projects and foreign trade flows was also considered by the authors of this paper - Nekhoroshkov et al. [16].

In general, despite its obvious relevance, the development of China, the Republic of Korea, Mongolia, and Russia and their integration interaction within the GTI are not fully reflected in the above research activities. At the same time, counting on the fundamental importance of the transport aspect of GTI interaction, one of the most vital issues is the assessment of foreign trade flows between the participating countries, those are representing the current and potential cargo base of transport projects. The first important step is the identification of specific areas for transport corridors development and the implementation of prioritized projects for them. Every project should be focused on reducing the influence of border barriers and the transport infrastructure improvement in order to meet the demands of regional economic integration, enabling the sustainability and expediency of foreign economic cooperation between the GTI member countries. The purpose of this paper is to assess foreign trade flows between GTI member countries and to define a cargo base for transport and logistics infrastructure projects development for the entire Greater Tumen region.

2 Materials and Methods

It is proposed to assess the foreign trade flows of the GTI member countries as a current and potential cargo base for transport projects by two key points - stability and complementarity, that will reflect the real state in current conditions and prospects for the development of foreign economic cooperation within the GTI.

We propose to assess the stability of foreign trade flows as a cargo base (current and potential) of transport projects within the GTI by absolute time series, which reflects the trend (or its absence, i.e. each next level is higher or lower than the previous one) and its stability, as well as by relative time series, i.e. growth rates, which reflects the stability of the change rate (the growth rate of each subsequent period is higher or lower than the previous one). For this, it is necessary to take the formula for calculating the Spearman coefficient (SC) as a basis:

$$SC = \frac{12 \times \sum_{i=1}^{n} M_i}{n^3 - n} - \frac{3 \times (n+1)}{n-1} \tag{1}$$

where:

SC - stability of the trend/rate of foreign trade cooperation,
M_i - product of numbers of time periods and their corresponding ranks of absolute/relative values of export/import,
n - duration of observation period.

The coefficient belongs to the range $[-1; +1]$ and has the following quality characteristics: less than 0.2 - instability; [0.2; 0.4] - weak stability; [0.4; 0.6] - average stability; greater than 0.6 - high stability. This assessment, by comparing the stability of absolute and relative indicators, will significantly reveal the areas of changes in the dynamics of cooperation between the GTI member countries.

Based on the assessment and analysis of the foreign trade cooperation dynamics stability within the GTI areas, it seems possible to identify emerging crisis phenomena and zones of its safe development. In accordance with the level of stability and dynamics trend direction for absolute indicators of exports and imports, along with the stability of the slowdown or increase in the rate of change, points are awarded for each pair of GTI member countries, reflecting the strength of the crisis state sign (signal) (Table 1).

Table 1. A set of signals about crisis phenomena in foreign trade cooperation of the GTI countries.

Stability of the rate		Stability of the trend					
		Weak stability		Medium stability		High stability	
		+	–	+	–	+	–
Weak stability	+	0	3	0	3	0	4
	–	2	3	1	3	0	5
Medium stability	+	0	3	0	3	0	4
	–	2	4	1	4	0	5
High stability	+	0	3	0	3	0	4
	–	3	4	2	4	1	5

After assigning points for the areas of foreign trade flows within the GTI, it is necessary to calculate the indicator of true conditions, i.e. the number of areas with a signal strength from 1 to 5, and an indicator of the total signal strength. These indicators are used to calculate the scale and intensity of the signals.

The scale reflects the breadth of crisis coverage of foreign trade cooperation within the GTI:

$$M = \frac{S}{\sum_{j=1}^{m} n_j} \times 100\% \tag{2}$$

where:

M - integral signal scale indicator,
S - true signal condition indicator,
n_j - total number of areas of foreign trade cooperation.

The intensity reflects the depth of crisis coverage of foreign trade cooperation within the GTI:

$$I = \frac{F}{\sum_{j=1}^{m} n_j \times r} \times 100\% \tag{3}$$

where:

I - integral indicator of signal intensity,
F - indicator of cumulative signal strength,
n_j - total number of areas of foreign trade cooperation,
r - dimension of the scale of the numerical values of signals.

Using the indicators of the scale and intensity of signals, it seems possible to recognize the stages of crisis phenomena in foreign trade cooperation in key areas of the GTI, that forms the basis of the potential transport congestion within its framework. A visual representation for the possible results interpretation is presented in Table 2.

The potential for foreign economic cooperation of the GTI member countries seems to be possible to assess using the complementarity index. This index is an indicator of the practicability of cooperation, and it allows to assess the extent of the commodity and sectoral differentiation of national producers and makes it possible to organize complementary technological and production ties. The higher the value of this index, the more countries can realize their benefits in trade through mutual compensation of needs based on trade specialization. From the point of view of foreign economic cooperation, this indicator characterizes the reserve of optimization of the national economic complex of each GTI member country and the possibility of increasing its efficiency.

Table 2. Recognition of the stages of crisis phenomena in foreign trade cooperation of GTI: the scale and intensity of signals [%].

Scale (S)	Intensity (I)	Crisis stage
0–30	0–30	None (green zone)
30–50	0–30	Potential (green zone)
30–50	30–50	Hidden nascent (green zone)
50–60	0–30	Hidden nascent (green zone)
50–60	30–50	Hidden developing (green area)
50–60	50–60	Hidden developing (yellow zone)
60–70	0–30	Hidden nascent (green zone)
60–70	30–50	Hidden developing (yellow zone)
60–70	50–60	Hidden progressive (yellow zone)
60–70	60–70	Hidden progressive (yellow zone)
70–90	0–30	Hidden developing (yellow zone)
70–90	30–50	Hidden progressive (yellow zone)
70–90	50–60	Hidden progressive (yellow zone)
70–90	60–70	Progressive turning into acute (red zone)
70–90	70–90	Progressive turning into acute (red zone)
90–100	0–30	Hidden developing (yellow zone)
90–100	30–50	Hidden progressive (yellow zone)
90–100	50–60	Hidden progressive (yellow zone)
90–100	60–70	Progressive turning into acute (red zone)
90–100	70–90	Sharp (red zone)
90–100	90–100	Sharp (red zone)

$$S_{ejmk} = 1 - \frac{\sum_{i=1}^{n} \left| E_{ij} - M_{ik} \right|}{2} \tag{4}$$

where:

S_{ejmk} - complementarity index,
j - exporter,
k - importer,
i - goods,
E_{ij} - part of goods i in the total exports of country j to the world market,
M_{ik} - part of goods i in the total import of country k from the world market.

3 Results

Over the past five years, the dynamics of foreign trade flows defined as a cargo base for transport projects within the GTI showed positive trends, with the exception of certain directions and time periods (Table 3). The increase in the final indicators was observed in all directions, except for the export of the Republic of Korea to China, Mongolia to Russia and the Republic of Korea. The average share of the GTI in the export of participating countries is more than 11%.

Table 3. Dynamics of foreign trade cooperation of the GTI member countries, USD billion.

Direction	2015	2016	2017	2018	2019
China-Republic of Korea	101.47	94.66	102.83	109.03	110.98
China-Russia	34.81	37.51	43.15	48.01	49.48
China-Mongolia	1.57	0.99	1.26	1.65	1.83
Republic of Korea-China	137.14	124.43	142.12	162.16	136.21
Republic of Korea-Russia	4.69	4.77	6.91	7.32	7.78
Republic of Korea-Mongolia	0.25	0.21	0.23	0.31	0.29
Mongolia-China	3.90	3.88	5.27	6.51	6.77
Mongolia-Russia	0.08	0.06	0.07	0.09	0.07
Mongolia-Republic of Korea	0.07	0.01	0.01	0.02	0.03
Russia-Republic of Korea	13.09	10.03	12.35	17.83	16.36
Russia-China	27.31	28.02	38.92	56.04	56.79
Russia-Mongolia	1.08	0.90	1.33	1.61	1.73

The total share of GTI directions in the export of the Republic of Korea is almost 26.6%, of which more than 25% falls on export flows to China and about 1.5% - to Russia. Note that the share of Russia has consistently increased over the period under review. Mongolia, as another GTI direction, has a fairly stable share, although its share in the total export of the Republic of Korea is insignificant (0.05%). In absolute terms, the export of the Republic of Korea to China was characterized by multidirectional dynamics, with the value of 136.21 billion dollars at the end of the period under review. The average growth rate at the level of 100.8% was caused by significant reductions in the value of export flows in 2016 (by 9.2%) and in 2019 (almost 16%), while in 2017–2018, the growth was at the level of 14.1–14.2%. According to the results of the period, the export from the Republic of Korea to China decreased by slightly less than 1%. The export of the Republic of Korea to Russia as a GTI direction was characterized by an annual increase, and as a result the final figure reached $7.78 billion, that is 65.9% higher than the level at the beginning of the period. The average growth rate was 114.7%. From 2017 to 2019, the value of export flows from the Republic of Korea to Mongolia has steadily increased, although the absolute figure is insignificant. A 5.4% reduction in 2019 led to the absolute value of exports at $0.29 billion, which is still 18.2% higher than at the beginning of the period, although it is noticeable only at the hundredth level.

In 2019, the total share of GTI directions in Russian exports was 17.71%, and it was the maximum value for the entire period. This stable dynamics is explained by the consistent increase in the share of China to 13.43% and Mongolia to 0.41%. The Republic of Korea reduced its position in the export structure to 3.87%, but this value is not the minimum for the period. According to the results of the period, the value of Russia's export flows to China reached $ 56.79 billion, thereby increasing more than 2 times compared to the initial indicator. The dynamics of exports was characterized by an extremely positive trend, while a significant increase was recorded in 2017–2018 by 38.9–44.0%. In other time periods, the increase was approximately from 1.5% to 2.5%. Russian exports to the Republic of Korea also showed an increase in the period by 24.9% from $ 13.09 billion to $ 16.36 billion. The average growth rate was at 121.7%, the dynamics was characterized by being in the positive zone, for excluding 2017 (a 23.4% decrease) and 2019 (an 8.3% decrease). Since 2017, Russia's export flows to Mongolia have shown positive dynamics with an average growth rate of 114.9%. As a result, the export value reached $1.73 billion, that is 60% higher than at the beginning of the period.

Chinese exports in the GTI directions are 6.5% in total, which is the maximum indicators for the period 2015–2019. The largest share among the GTI member countries in China's exports belongs to the Republic of Korea - 4.4%, the share of Russia is about 2%, less than 0.1% falls on Mongolia. At the end of the period, the value of China's export flows to the Republic of Korea increased by 9.4% to $ 110.98 billion. Note that after a decrease in the indicator by almost 7% in 2016, there was a stable increase at the level from 2% to 8% in subsequent years. China's exports to Russia were characterized by an annual average growth rate of 9.3%. In absolute terms, the final figure was $ 49.48 billion, which is 42.2% higher than the same indicator at the beginning of the period. Export flows to Mongolia showed stable positive dynamics after declining by almost 37% in 2017, but their growth varied significantly - from 11.1% in 2019 to 30.6% in 2018. In general, the final absolute value was at the level of $ 1.83 billion - an increase over the period of 16.4%.

The share of GTI directions in Mongolia's export flows is 90.1%, while the average value for the period is 87.5%. The largest share traditionally belongs to China - almost 89% at the end of 2019, the shares of Russia and the Republic of Korea in this structure are 0.9% and 0.4%, respectively. At the end of the period, the value of Mongolia's exports increased only towards China - by almost 74% (to $ 6.77 billion), which was preceded by a stable positive trend (an average increase of 15.7%), except for an extremely insignificant reduction in 2017. Mongolia's exports to Russia decreased by 11.4%, although due to the positive trend in 2017–2018, the average growth rate is at 100%. In general, the absolute values of the indicators are insignificant - $ 0.07 billion in 2019. An even more significant decline was observed in exports to the Republic of Korea - by 58.2% to the final figure of $ 0.03 billion. At the same time, over the past three years, the value of exports has steadily increased. However, the 2017 decline by 87.2% and was not compensated.

The results of calculating the stability coefficient of foreign trade cooperation within the GTI (Table 4) indicate that the export of China to the Republic of Korea, Mongolia to China, Russia to Mongolia is characterized by a highly stable positive trend, while the rate of positive changes slows down with weak stability. A similar trend and the degree of its stability are typical for Russian exports to China. However, the rate of positive changes slows down with weak stability, as well as for Chinese exports to Mongolia, the Republic of Korea in Russia and Mongolia, while positive changes are increasing their rate with medium stability. China's exports to Russia are also highly stable and positive in their trend, but with a moderately stable deceleration. Russia's exports to the Republic of Korea are characterized by an average stability of the positive dynamics of absolute indicators, and the rate of these flows increases with a high degree of stability. Weak stability of dynamics was recorded for the exports of Mongolia to Russia and the Republic of Korea to China. Moreover, in the first case, an increase in the rate of positive changes with an average stability is observed, in the second - a slowdown in the rate with the same degree.

Table 4. Diagnostic matrix of stability of foreign trade cooperation within the GTI ("+" - positive dynamics, "–" - negative dynamics).

Stability of relative indicators (growth rates), stability of rate		Stability of absolute indicators, stability of the trend							
		K < 0.2, instability		0.2 < K < 0.4, weak stability		0.4 < K < 0.6, medium stability		K > 0.6, high stability	
		+	–	+	–	+	–	+	–
K < 0.2, instability	+	–	–	–	–	–	–	–	–
	–	–	–	–	–	–	–	–	–
0.2 < K < 0.4, weak stability	+	–	–	–	–	–		China-Korea, Mongolia-China, Russia-Mongolia	–
	–	–	–	–	–	–	–	Russia-China	–
0.4 < K < 0.6, medium stability	+	Mongolia-Korea	–	Mongolia-Russia	–	–	–	China-Mongolia, Korea-Russia, Korea-Mongolia	–
	–	–	–	Korea-China	–	–	–	China-Russia	–
K > 0.6, high stability	+	–	–	–	–	Russia-Korea	–	–	–
	–	–	–	–	–	–	–	–	–

Based on the presented calculation method, the indicators of the scale and intensity of crisis signals are less than 10%, which indicates the absence of crisis phenomena in foreign trade cooperation between the GTI member countries, and the zone of its development is green. In other words, the stability of trends and the rate of development of foreign trade cooperation ensure the highest level of security so as not to go into crisis stages, even considering the unfavorable situation in the world economy. The assessment of the feasibility of the development and implementation of transport projects for various long-term scenarios should be associated with the presented assessment and the security zone for the development of foreign trade cooperation identified and based on its results, thus this cooperation reflects the key areas those provide the cargo base of GTI projects.

On the other hand, in addition to the stability of the existing foreign trade flows within the GTI, a fundamental point is to identify the potential for cooperation between the participating countries, i.e. in fact, the expediency of these flows. The complementarity indexes were calculated for pairs of countries (Table 5).

Table 5. Indices of complementarity of foreign economic cooperation within the GTI.

Exporter	Importer			
	Russia	Republic of Korea	China	Mongolia
Russia	–	0.34	0.35	0.48
Republic of Korea	0.57	–	0.63	0.67
China	0.69	0.64	–	0.75
Mongolia	0.10	0.11	0.28	–

The highest complementarity index reflects China's export flows to Mongolia and Russia, as well as the Republic of Korea to Mongolia. Mongolia's exports in all directions within the GTI are characterized by the least complementarity. The average value of the complementarity index within the GTI is about 0.5 (50%), which indicates not only the potential of foreign economic cooperation for member countries, but also the need to take measures to further intensifying, including foreign trade transport services improvement.

4 Discussion

The principled character of a comprehensive and timely study of the issues of substantiation of various transport projects is explained by the fact that transport is the most significant and feasible area of cooperation within the GTI. Note is taken that the original Tumen project (1990) was mainly "a transport hub project". As for the GTI, several transport problems were solved in the first few years, including the implementation of a project for an international ferry service between the ports of Zarubino (Russia), Sokcho (Republic of Korea) and Niigata (Japan).

Currently, there are several key transport projects within the GTI:

- creation of vehicle ferry system in North-East Asia,
- modernization of the port of Zarubino,
- completion of the construction of the Makhalino (Kraskino) - Hunchun railway line and the beginning of a full-fledged transport interaction along this route,
- a road and harbor construction project at China and North Korea border.

Note that some of these projects are not implemented, and even are not planned for implementation due to a number of reasons. In August 2013, at a meeting of the GTI Transport Council in Vladivostok, its participants approved a comprehensive study of intermodal routes in Northeast Asia, carried out by experts (China, Mongolia, Russia, South Korea - an assessment of the current economic demand for transport), "Regional transport strategy and GTI action plan". In accordance with the developed concept, the transport infrastructure of Northeast Asia is represented by six corridors.

The first corridor includes the ports of Zarubino - Posiet - Rajin, passes through the territories of Hunchun - Changchun - Orksan - Eastern Mongolia, and goes to the Trans-Mongolian railway of the Trans-Siberian Railway. The second one runs to the north and can also give Mongolia the opportunity to enter Northeast Asia. This is the "Suifenhe" transport corridor, connecting the port of Vostochny (Nakhodka) - through Grodekovo - Pogranichny - Suifenhe, Xinjiang province - Inner Mongolia with access to the Trans-Siberian Railway. The main purpose of this corridor is to enhance the transport potential of this direction. The third corridor is a fragment of the Trans-Siberian Railway connecting the territories of the Trans-Baikal region and the Primorsky Territory. The fourth transport corridor - "Dalian" - unites sections along the route from Blagoveshchensk, Heihe to a group of ports in the Dalian region. The fifth corridor is the western Korean one, running through the Korean Peninsula from Busan to Seoul, to Shenyang - Harbin in China. The sixth is the East Korean corridor, passing through the port of Rajin in North Korea, with an exit to the Trans-Siberian Railway.

According to preliminary estimates, the initial investment that will be required to push the development of GTI corridors is more than $ 3.5 billion. These calculations included a number of projects: an updated feasibility study (FS) of the container terminal project in Zarubino, feasibility study of projects in Eastern Mongolia, the construction of logistics centers in Hunchun, etc. Regarding transit container transport, it is necessary to simplify the procedures for their movement across borders. It should be noted that the Asian Bank for Reconstruction and Development, the Export-Import Bank of China and Vnesheconombank are ready to support the strategy within the GTI. It should be noted that transport cooperation within the GTI has technical difficulties associated with the different gauge of the railway track in Russia and other participating countries.

The regional transport strategy has become the most important tool for the implementation of transport goals within the GTI. Its key goal was to develop a reliable, cost-effective and efficient integrated transport network by planning and promoting the activation and development of international transport corridors. The strategy was clearly structured in areas, within each of which action plans were defined:

- transport connectivity,
- support to the improvement of transport infrastructure,
- transport corridor operation software,
- transport corridor management,
- private sector participation.

Unfortunately, it cannot be argued that the real results of this strategy have been achieved. But this does not mean that transport issues have remained outside the agenda of the GTI member countries, they are still being discussed and planned. For example, in accordance with the Seoul Declaration (December 16, 2020), the development of the GTI transport sector involves ensuring the smooth operation of transport corridors in Northeast Asia by harmonizing and simplifying border crossing formalities and removing existing barriers for intermodal transit traffic. Note that the Russian side just focuses on transit routes, which can become the basis for the implementation of the strategy of transport corridors within the GTI. Indeed, the use of Russia's transit potential may become a fundamentally new area of transport cooperation. We are talking about improving the Eurasian transport corridors and routes (modernization of the Trans-Siberian railway, the Baikal-Amur railway, Pacific ports, etc.), the development of new routes providing access from Europe to the ports of the Russian Far East and further to the countries of Northeast Asia.

However, taking into account the prevailing trends in Europe-Asia transportation, it is not entirely correct to make the implementation of GTI transport projects dependent on prospective transit freight flows. At first, it is necessary to concentrate on the transport support of foreign trade flows directly between the GTI member countries. The results of the assessment of foreign trade cooperation indicate a positive trend in all areas, while the degree of its stability varies from weak to high. In the overwhelming number of directions, both a high stability of the trend and a high stability of the increase in the rate of positive changes are observed. All this testifies to the high level of foreign trade cooperation within the GTI, which, from these positions, is developing in a safe zone with no signs of crisis phenomena. Besides, the member countries have the potential to build up foreign trade relations, allowing them to complement technological production ties in accordance with the commodity diversification of exports and imports.

Consequently, improving the transport and logistics infrastructure in the Tumen region is of a vital necessity in accordance with the existing foreign economic relations between the GTI member countries, their current state, and prospects for expansion. The existing cargo base can influence the efficiency of transport projects in the region, and on the other hand, the advanced development of the transport sector can become a real trigger for economic cooperation and the transformation of GTI into an independent intergovernmental organization.

5 Conclusions

In general, it can be concluded that the GTI member countries have significantly increased their potential for intergovernmental cooperation to promote trade relations in the region. Foreign economic cooperation within the GTI is obviously the basis for interaction in various fields, especially in transport. On the other hand, the development of some transport corridors and improvements in border crossing processes, transshipment points, etc. are made directly dependent on their congestion with foreign trade flows. Consequently, the development and feasibility of implementing transport projects of GTI should be based on the potential of foreign economic cooperation between its member countries.

Based on the results of the assessment of foreign trade flows between the member countries, that forms the current and potential cargo base of transport projects, it was concluded not only about a high degree of stability and potential of foreign economic cooperation, but also about the need of extra measures for its further intensifying, including foreign trade transport services improvement.

In this context, transport projects should be focused on improving transport and logistics infrastructure, simplifying border-crossing procedures, and developing improved types of transport and logistics services and technologies. There is a need to implement the tasks of transport connectivity through existing transport corridors, which requires an improvement in their performance characteristics, as well as the construction of new corridors. For more efficient services in foreign trade flows, and transit in the future, it is necessary to develop interstate agreements in the field of transport, the implementation of those would benefit to all GTI member countries. The transport corridor management system is also of fundamental importance, including coordinating organizations created by the member countries together.

References

1. Lu, S., Shen, J.H., Li, W., Zhang, J.: A theory of economic development and dynamics of Chinese economy. Econ. Model. **86**, 69–87 (2020). https://doi.org/10.1016/j.econmod.2019.06.003
2. He, L.Y., Lin, X., Zhang, Z.: The impact of de-globalization on China's economic transformation: evidence from manufacturing export. J. Policy Mod. **42**, 628–660 (2020). https://doi.org/10.1016/j.jpolmod.2020.02.001
3. Fleisher, B.M., Zhang, X.: China economic review 30 years. China Econ. Rev. (2021). https://doi.org/10.1016/j.chieco.2021.101616.. ISSN 1043-951X
4. Sim, S.G., Yoo, D.: A quantitative study on endogenous formation of comparative advantage in South Korea. Int. Rev. Econ. Finance **53**, 71–87 (2018). https://doi.org/10.1016/j.iref.2017.10.005
5. Cheng, C.H.J.: Effects of foreign and domestic economic policy uncertainty shocks on South Korea. J. Asian Econ. **51**, 1–11 (2017). https://doi.org/10.1016/j.asieco.2017.05.001
6. Lim, E.S., Breuer, J.B.: Free trade agreements and market integration: evidence from South Korea. J. Int. Money Finance **90**, 241–256 (2019). https://doi.org/10.1016/j.jimonfin.2018.09.010

7. Chu, A.M.Y., Lv, Z., Wagner, N.F., Wong, W.K.: Linear and nonlinear growth determinants: the case of Mongolia and its connection to China. Emerg. Mark. Rev. **43**, 100693 (2020). https://doi.org/10.1016/j.ememar.2020.100693
8. Thorbecke, W.: East Asian value chains, exchange rates, and regional exchange rate arrangements. J. Asian Econ. **65**, 101132 (2019). https://doi.org/10.1016/j.asieco.2019.101132
9. Voskoboynikov, I.: Sources of long run economic growth in Russia before and after the global financial crisis. Russ. J. Econ. **3**, 348–365 (2017). https://doi.org/10.1016/j.ruje.2017.12.003
10. Mitsek, S.: Econometric model of Russian Federation: analysis and forecasts for 2018–2019. IFAC-PapersOnLine **51**, 395–400 (2018). https://doi.org/10.1016/j.ifacol.2018.11.416
11. Ono, S.: Financial development and economic growth nexus in Russia. Russ. J. Econ. **3**, 321–332 (2017). https://doi.org/10.1016/j.ruje.2017.09.006
12. Bayramov, V., Rustamli, N., Abbas, G.: Collateral damage: the Western sanctions on Russia and the evaluation of implications for Russia's post-communist neighbourhood. Int. Econ. **162**, 92–109 (2020). https://doi.org/10.1016/j.inteco.2020.01.002
13. Magazzino, C., Mele, M.: On the relationship between transportation infrastructure and economic development in China. Res. Transp. Econ. (2020). https://doi.org/10.1016/j.retrec.2020.100947
14. Foo, N., Lean, H., Salim, R.: The impact of China's one belt one road initiative on international trade in the ASEAN region. North Am. J. Econ. Finance **54**, 101089 (2019). https://doi.org/10.1016/j.najef.2019.101089
15. Lall, S.V., Lebrand, M.: Who wins, who loses? Understanding the spatially differentiated effects of the belt and road initiative. J. Dev. Econ. **146** (2020). https://doi.org/10.1016/j.jdeveco.2020.102496
16. Nekhoroshkov, V., Aroshidze, A., Nekhoroshkov, E.: Transport support of Europe-Asia trade: structural shifts. Transp. Res. Procedia **54**, 182–190 (2021). https://doi.org/10.1016/j.trpro.2021.02.063

Author Index

E. Macioszek and G. Sierpiński (Eds.): TSTP 2021, LNITI, pp. 201–202, 2022.
https://doi.org/10.1007/978-3-030-93370-8